SLOW COMPUTING
Why We Need Balanced
Digital Lives

Rob Kitchin and Alistair Fraser

BRISTOL
UNIVERSITY
PRESS

First published in Great Britain in 2020 by

Bristol University Press
University of Bristol
1-9 Old Park Hill
Bristol
BS2 8BB
UK
t: +44 (0)117 954 5940
e: bup-info@bristol.ac.uk
Details of international sales and distribution partners are available at
bristoluniversitypress.co.uk

British Library Cataloguing in Publication Data
A catalogue record for this book is available from the British Library.

ISBN 978-1-5292-1126-9 paperback
ISBN 978-1-5292-1128-3 ePub
ISBN 978-1-5292-1127-6 ePdf

Cover design and cover image credit: Liron Gilenberg
Printed and bound in Great Britain by TJ International,
Padstow
Policy Press uses environmentally responsible print
partners.

In loving memory of

Mervyn Frank Kitchin (1944–2017)

and

Agnes McCready Stewart (1941–2019)

Contents

About the authors

Alistair Fraser is Lecturer in the Department of Geography at Maynooth University, Ireland. His research focuses on the interaction of agrarian change and the food economy, based on research fieldwork in South Africa, Uganda, and Mexico. He is now conducting research on the way digital technologies, such as precision agriculture and curated social media, alter the geographies of food. He is the author of *Global Foodscapes: Oppression and Resistance in the Life of Food* (Routledge, 2016) and numerous articles in academic journals, and is an editor of the journal *Human Geography*. https://www.maynoothuniversity.ie/people/alistair-fraser

Rob Kitchin is Professor in the Maynooth University Social Sciences Institute, Ireland. He wrote his first article about the internet in 1995 and has conducted extensive research on digital technologies and their impact on society. He is (co)author or (co) editor of 30 non-fiction books, including *Mapping Cyberspace* (Routledge, 2000), *Code/Space: Software and Everyday Life* (MIT Press, 2011), *The Data Revolution* (Sage, 2014), *Understanding Spatial Media* (Sage, 2017), *Digital Geographies* (Sage, 2018), *The Right to the Smart City* (Emerald, 2019), and *How to Run a City Like Amazon, and Other Fables* (Meatspace Press, 2019). He has been an editor of three leading geography journals and editor-in-chief of the 12-volume *International Encyclopedia of Human Geography* (Elsevier, 2009). He is a recipient of the Royal Irish Academy's Gold Medal for the Social Sciences. https://www.maynoothuniversity.ie/people/rob-kitchin

Acknowledgements

Some of the research for this book was funded by an Advanced European Research Council grant, The Programmable City, ERC-2012-AdG-323636-SOFTCITY and a Science Foundation Ireland grant, Building City Dashboards, 15/IA/3090.

The genesis for this book was a working paper written for the Slow Computing workshop that we organized in Maynooth in December 2017. Many thanks to the attendees for their thoughtful papers and productive discussion. We are grateful to Mike Bratt, Mark Boyle, and the anonymous reviewers of the book proposal and manuscript for their helpful suggestions. Many thanks to Paul Stevens at Bristol University Press for enthusiastically supporting publication of the book, and the rest of the team at BUP for production and marketing.

1

Living Digital Lives

Monday. The week begins. It's 6.30am. You've been woken by your phone's alarm. You immediately reach for the device, noticing there are WhatsApp messages waiting to be read. An icon from Twitter is also alerting you to something, perhaps a new follower. Your partner has been up for a while. In the living room, Amazon's Alexa has begun streaming some songs from Spotify. You know you need to check email. Then check your calendar to see what meetings you have scheduled in the afternoon. You notice your 'bedtime' tablet, which you use to watch Netflix, is flashing to be charged. "There's something wrong with the screen on the fridge," your partner says. Time to get up.

This start to the day won't be exactly the same as yours. You'll have your own array of devices. You might not use Twitter. Maybe you live on your own. And maybe you don't (yet) have a screen on your fridge. But for many of us today, regardless of how we go about waking up, we soon start to connect and interact with digital devices and services. In fact, according to a 2017 report by Deloitte, 16% of Americans reach for their smartphone immediately after waking, 42% within five minutes, and 62% check for messages, email, social media, and the news within 15 minutes.[1]

We lead digital lives. Twenty-six per cent of Americans report being 'almost constantly' online.[2] Over 80% of people own a smartphone, with the average person checking it about 50 times a day.[3] A large proportion have it at hand at all times, even when asleep.[4] We often check in during the night; a 2017 survey in the UK found that 66% of teenagers wake to check their phone in the early hours, with 35% of parents doing the same.[5] And even if we do not own a smartphone, or use the internet infrequently, we still interface with digitally mediated services all the time – at work, moving through transport systems, using utilities, passing surveillance cameras, and so on.

Is this just the way life is now? Should it be like this? Does it matter at all?

In this book we make two related arguments. First, how we interact with digital devices *does* matter. Digital technologies are accelerating and fragmenting our everyday lives, and the data our devices gather are used to profile and target us. Second, we should step back, even if just a little, to try and seize some self-control. Our call is not a Luddite one. We are not against computing and the digital age. Rather, our call is for balance: for using today's computing power, but in a way that is managed, considered, and to our benefit. 'Slow computing' is one possible way you can be more careful about leading a digital life. But it is more than that. Slow computing is also about seeking and making changes to how our digital society and economy operate and are organized.

Slow computing is not an easy strategy, however. It is something to work at, albeit in the face of numerous pressures. Many digital technologies are designed to be seductive and addictive.[6] We're called upon, indeed often psychologically compelled, to check notifications on our phones, to answer emails, reply to messages, and see what's happening on social media, or encouraged to just jump lazily around from one app to another, perhaps casually tapping at the screen as we move from Netflix to YouTube and then Facebook. Devices draw us in: laptops desktops, tablets, phones, TVs, and yes, even fridges, demand our attention. They seduce us, making us willingly and voluntarily subscribe to and desire their logic, trading any potential negative effect, such as a loss of privacy or time fragmentation and stress, against perceived benefits such as enjoyment, convenience, and cost. We're also compelled to use digital devices to complete mundane tasks: interfacing with government; filing taxes; applying for a permit or benefits; filling in forms to update an insurance policy; buying goods or services; or looking for information such as a library's opening hours.

Participating in everyday life, therefore, brings us into contact with the digital world; and the experience of interacting with, and through, digital media often makes us want to do it more, even if it works against our self-interest. Moreover, states and corporations often have a lot of power to set and drive agendas; this means resisting and pushing back can be daunting and involve energy we might not always manage to spare. Against these pressures, slow computing seeks to redirect attention, reclaiming time for other pursuits, and to protect its practitioners from any pernicious effects of living a digital life. It does this both personally, by providing tactics for managing how you engage with a digitally mediated world, and collectively by pooling knowledge, cooperatively acting together, and mobilizing political power to shift public debate and influence the regulation of questionable practices.

In this book we reveal why we need *balanced* digital lives. We demonstrate why slow computing deserves our attention and detail how we can individually and collectively make it achievable. We make clear the issues at stake, set out slow computing thinking and practices, and document the practical and political interventions that we can undertake to reassert sovereignty over our digital lives. Hopefully, the thesis we develop and the strategy outlined will convince you to join the countless others who are already starting to practise slow computing as an alternative means to experience the joys of computing.

The joys of computing

The quality and power of today's digital devices and services should not be understated. Many of us hold in our hands extraordinary equipment. On our smartphones we can communicate with satellites to identify our location; download and watch movies when we're out and about; send photos to friends on the other side of the world; make video calls; play games; attach other devices to measure weights or temperatures, or take credit card payments; use default apps such as a calculator, a calendar, a note-taker or an interactive map; and download and use thousands of other apps for exercise, cooking, reading, weather, travel, radio and podcasts, wellness advice, and just about anything else you can think of. Computers are generative technologies par excellence – they enable us to creatively and flexibly produce all manner of designs, stories, and products in ways that far exceed the capacities of previous tools. They provide diverse forms of entertainment through a single device and allow us to pursue multiple forms of communication, networking, and consumption. They are truly remarkable devices.

Lots of us would be quite surprised (and distressed) if a device stopped working. It would be unusual for the screen to break without us having dropped it. Software programs do sometimes crash these days, but a great many work like clockwork. Batteries are not always replaceable but many last quite well, even if they do seem to need charging every day. And as digital devices become ever more robust, with powerful processors, lots of memory, and high-definition screens, we use them to mediate and augment everyday life: to produce, consume, interact, communicate, govern, police, travel, and play. And we can do this while on the move – checking a bank account, sending an email, watching live sport, or following a recipe is becoming easier and quicker.

We are, in short, living in an age of unprecedented computing power, with some indications that even people living in very remote places or who belong to the world's poorest communities will soon have devices with regular, fast, and affordable internet connections. In this sense, we are living at the start of the age of what some call 'ubiquitous' (computation available everywhere, via devices able to connect to the internet) and 'pervasive' (computation embedded in everything enabling them to be linked to the internet) computing.[7] Such are the capacities and utility of digital technologies that they have become the lifeblood of today's information society and economy, just as steam was at the start of the industrial age.[8] Digital computation, like steam, is reshaping our world – from automating manufacturing and service jobs, to making cashless payments, optimizing the delivery of utilities, creating autonomous vehicles and drones, monitoring personal health, sharing information and viewpoints, and remotely controlling home appliances.

The speed, power, and stability of many of today's devices and services is something to celebrate. We certainly don't want to go back to an analogue world, and we doubt you do, either. For all we might complain or worry about the way we now live with digital devices and services, few of us want to return to the pre-digital age. We enjoy our devices – we even find joy (not to mention love, good times, friendships, work, and crucial information) in them. To labour the point a little more here, just think of some of the other obvious conveniences or subtle helping hands provided by today's technologies:

- The train you took this morning had free Wi-Fi. It wasn't lightning fast but you were able to update your newspaper app and pass the time.
- While at work, your employer emailed you a message to say this month's payslip is available for you to review. You saw you've been refunded the expenses from that trip abroad last month. Doing that took about 45 seconds.
- In your local supermarket you simply pick items, scan them, place them in your bag, check out, and leave. No waiting in line. No unpacking/repacking. You're in and back out again in less than five minutes.
- Your child's school has sent a convenient reminder on Facebook about next Friday's theatre production.
- Your car's GPS was updated over-the-air today and on the way to a friend's house it rerouted you to avoid a nasty traffic accident near the church.
- At work or at home you've been able to simultaneously edit a report with other contributors who are located in another office.
- The new office chair you ordered online yesterday has arrived. It came at the same time as a new 256GB Micro SD card for your camera, which lets you take and store thousands of photos.
- Using your phone, you follow the actions, thoughts, and adventures of your friends, as well as keeping up to date with breaking news on social media.
- You compare dozens of prices and book a hotel; then pay a utility bill instantly from your armchair.
- Using a networked games machine, you play a multi-player game in real-time with several friends who are all located in different countries.

You'll be familiar with all these features of everyday life, even if the contexts or specifics are different. Like us, then, you probably have to admit that there's some joy to be had through networked computation. In many ways, digital technologies do simplify things. The weight of some everyday problems has been reduced. A sense of convenience is created. Time and energy is seemingly saved. To make it absolutely clear, there's a lot to like about the way today's world of technology is taking shape.

Problems coming into focus

For all the joys of computing, however, it's also true that collectively as a society we are becoming aware of some major problems with the way today's digital technologies are configured, rolled out, and utilized. Some of these problems are personal and some are institutional and structural.

Lots of us feel tethered to and bogged down by devices. We feel harried, rushed, and constantly pestered by messages arriving and needing a response. We feel a compulsive need to check for updates, to interact, or to continue playing a game. Indeed, we sometimes seem to be addicted to digital life; in fact, we *are* addicted. We often seem unable to function without having at least one device within reach and giving the sensation that some form of salvation is only a quick click away. Indeed, 31% of Americans feel anxious when separated from their phone, with 60% experiencing occasional stress when their phone is off or out of reach.[9] In just a few years, it seems as if we have become beholden and tethered to digital devices and systems.

This tethering is compounded by societal expectation from family, friends, employers, clients, service providers, marketers, and others who expect that we are always reachable, that we will respond immediately, and that every part of our daily lives should be mediated by computation and connectivity. Homes are becoming ever more 'smart': televisions, washing machines, fridges, coffee machines, heating systems, and a plethora of other devices, are now connected to the internet and they actively try to learn and react to our preferences and respond to our requests. Work requires using ordinary computers, along with specialist devices, tailored software, and interactions and transactions across the internet. Interfacing with government requires filling in online forms and uploading information. An industry like air travel requires us to interface with digital computation at every stage: buying tickets, checking in, passing through security and immigration controls, routing baggage, and then flying in a plane that is reliant on millions of lines of code to take to the skies.[10] Indeed, such is the ubiquitous and pervasive extent of computing that it is now almost impossible to live outside its orbit. We are compelled to be digital citizens, whether we like it or not, almost regardless of whether we possess sufficient digital literacy and skills.

A key point to begin emphasizing here is that, at every point in these overlapping digital ecosystems, data are being generated, extracted, and translated into value. Websites and apps monitor what links we click on; large swathes of data concerning usage and location are extracted through apps; and our faces, vehicle licence plates, and the unique signatures of our devices (such as the MAC [media access control] address which identifies each smartphone) are scanned in public space and tracked. These streams constitute what has been termed 'big data': that is, data which are continuously produced in real-time, are exhaustive to a system (the data relate to *every* person, object, transaction within a system[11]), and uniquely identifiable at the individual level.[12] At the same time, there are technologies that collate and process these data to make inferences and predictions about what kinds of people we are and then use these insights to make all kinds of decisions about, and indeed for, us. For example, some retailers adjust prices according to who they think we are and what algorithms suggest we can afford to pay. They try to nudge us into making purchases we were not initially intending. Some platforms fire annoying, but sometimes also knowing, ads at us; some ads are even produced by political 'bots' looking to sway our opinions, often in unaccountable ways. Data brokers use the profiles to help companies make important life-changing decisions about whether we might be a suitable employee or tenant, or are sufficiently low-risk to loan money to, or who might be a suitable date for us. Extracted data are then employed to construct personal profiles that are then used to socially sort people into various categories, who are then treated differently.[13] Our data are being monetized and their value extracted, and used to reshape how states govern society. Our life chances are now mediated by forms of algorithmic governance, and in some cases life-or-death decisions can be made about us, sometimes autonomously and based on inaccurate data, such as by diagnostic healthcare technologies or killer drones.[14]

Consequently, how we understand and experience privacy is changing. Privacy is a condition that many people value and it is considered a basic human right in most jurisdictions, enshrined in national and supra-national laws in various ways. In the pre-digital age, citizens in western democratic countries expected a high degree of privacy – to be able to selectively reveal themselves to the world within accepted limits (such as the state's right to know certain information). The production of big data, however, challenges privacy rights because of the extensive, invasive, and granular nature of data extraction, and how the data are used for purposes for which they were not generated.[15] Big data and associated data analytics and machine learning also multiply the extent to which unacceptable practices and privacy harms – such as interrogation, identification, secondary use, exclusion, breach of confidentiality, disclosure, exposure, blackmail, appropriation, distortion, intrusion, decisional interference – can occur,[16] as well as creating new forms of predictive privacy harms through inferencing.[17] In other words, the proliferation of networked digital devices has fundamentally changed the privacy landscape and the extent to which citizens can expect to selectively reveal themselves. New legislation, such as the General Data Protection Regulations (GDPR) in the European Union (EU), are attempts to more effectually reassert privacy rights, though their effectiveness in this endeavour has yet to be fully assessed.

A multibillion dollar, global industry, the data brokerage firms that extract value from big data are now a central, if largely hidden, part of everyday life. As is the games industry, the sharing economy, digitally mediated retail, software development, apps and operating systems, web design, hardware manufacture, networking and telecommunications, and so on. We live in the age of what some initially termed the 'information economy'[18] and more recently 'surveillance capitalism'[19] or 'platform capitalism'.[20] Networked computation is underpinning and driving economic development, how the global economy is organized and functions, and how value is produced, extracted, and reinvested (even in long-standing traditional industries, such as manufacturing, logistics, agriculture, media, and tourism). In turn, the largest firms producing many of the digital goods and services we use are accumulating enormous profits, with huge market capitalizations created in a very short space of time. Google, Facebook, Uber, Alibaba, Amazon, Apple, and hundreds of others, are all young companies that have struck gold by disrupting old industries and creating new products and markets, quickly becoming globe-spanning enterprises. In turn, these companies wield enormous economic power and also exercise political power, lobbying governments and kicking back against regulations that might limit how they produce value and their ability to expand further. And yet, their wealth is often generated in large part through our data and labour, and infringing on our rights and expectations concerning privacy and data use.

Of course, it is not simply companies that seek to collate and extract value from data. Government at all levels (local, national, supra-national) generate and process data for the purposes of public administration, managing operations, and policing and security. In part, this is to ensure that citizens comply with laws and regulations (attending school, paying taxes, obeying traffic rules), but also receive their entitlements (to services, welfare, housing, health, and so on). More and more interactions with the state are conducted online, with citizens directly interfacing with government services and databases. Analytics are applied to these databases to identify potential welfare cheats and reduce fraud. Such systems also work to monitor the performance of government itself, with data used to assess the efficiency and effectiveness of programmes and policies, and to design new ways of delivering services. On a darker note, as the revelations of Wikileaks, Edward Snowden, and other whistleblowers have demonstrated, there also has been a step-change in the extent and nature of state-led surveillance in many nations (for example, the various programmes of the US National Security Agency [NSA] and UK Government Communications Headquarters [GCHQ], and many police forces). As well as parsing government data to assess potential security and criminal threats, these agencies are also drawing on commercial data to monitor the views (via social media, email, messaging, phone conversations), associations (via social networks), activities, and locations of populations.

What this all means is that, in living our digital lives, we can be exposed to some worrying practices. Even if your, or our, exposure might not always *seem* so bad, we are often unaware of the extent to which we constitute the product from which value is being leveraged, or how we are being negatively affected in overt and covert ways, or how others might be more profoundly affected.

Among university researchers, media critics, political commentators, and many business experts, there is a growing call for a better understanding of these emerging problems. In academic journal articles, scholarly books, newspaper or magazine exposés, and all manner of podcasts, concern is raised about the effects of digital life; about the ways in which disruption of established industries or government practices is not always a good thing; about how entire technical systems might discriminate against or disadvantage certain populations; or how new platform economies can destabilize labour markets and create more precarious work. Indeed, there have been sustained calls for more investment in establishing ethical guidelines for machine learning and artificial intelligence, regulating data privacy and security, enacting data justice, and for instituting new labour laws, among others.

We think one way to cut through and make sense of all of the criticisms and alarms is to suggest that digital life is creating worrying problems along two principal axes. On the first axis, the issue is what we'll refer to as '*acceleration*': a sense that, for all the convenience of digital technologies, they also cut into our lives in ways that increase the pressure to always do more, always remain connected, always stay on alert, leaving us struggling – at times, really struggling – to switch off, relax, and not re-connect. Emailing eats into our holidays; sending messages on WhatsApp disturbs our family time; devices beep and ping when we're eating out; completing everyday tasks means going online, jumping between websites, clicking here and then there in a rush, with a sense of urgency and almost panic because we know it shouldn't take long but the internet's slow for some reason and we need to run for a bus, but this online form has an error and we can't submit it and … and … pressure builds. Digital life seems to compress and fragment our time.

If acceleration is on one axis, we think '*extraction*' needs to be on the other. As we have noted, as digital life has emerged it has become reliant on a business model that opportunistically mines data about all aspects of our lives, including our tastes, ideas, and thoughts. We enjoy the fact that many digital services and apps are free to use. They are not free to create and run, however, and it's impossible to cover costs or make a profit with no income. So *we* have become the product, as well as the consumer and user, and in the case of some platforms the producer (for example, we create the content that is consumed on social media). In this sense, there is an exchange of data and labour for a service.[21] Even so, this exchange can be lopsided, and the implications not entirely clear and transparent to the user. The exchange can lead to billions of dollars of profits flowing to the companies that make the most of these opportunities. And what this relationship means for all of us is still not at all clear. To repeat, it is us who produce the data, become the targets of tracking and surveillance taking place behind the scenes, and us who are worked on by algorithms and analytics designed to game us in some way, often autonomously and in many ways without regulation or oversight. We're caught up in a data grab.

When we take stock of acceleration and extraction, therefore, we're struck by the way individual users – you, us – are in the centre of things with serious pressures bearing down from two directions. But we also know that individuals can choose, to varying degrees, how they act in response to acceleration and extraction, and can also work with others to counter these pressures. It is precisely this sense of acceleration and extraction inviting a response that we explore in this book. We're convinced everyone can and probably should work to find a balanced digital life; that the joy of computing exposes us all to acceleration and extraction, but also that we have options to establish new digital selves. The way we respond now and in the future is the core issue.

Slow computing

Slow computing is a way to characterize a type of response to digital life that prioritizes your needs and interests, as well as the public good for society as a whole. The term was introduced by Nathan Schneider in 2015 in a short article in *New Republic*.[22] He used it to refer to using a computer that has been self-configured with open source software. In a related article in *America* magazine he describes slow computing thus:

> What I mean by this is making choices about using computers, and the networks they connect to, with more awareness of how they affect ourselves and others around us. Much as the Slow Food movement emphasizes local economies, traditional knowledge and

ecology, Slow Computing means not merely opting for the most competitive, profit-driven hardware and software, but instead building a commons. It means cultivating digital lives that reflect our analog values.[23]

Schneider explicitly links slow computing to the slow living movement. Since the mid-1980s, this movement has been growing, with the ideas and values of 'slowness' being applied to different facets of everyday life and work.[24] So we can find people calling for and practising 'slow living' in general,[25] 'slow food',[26] 'slow scholarship',[27] 'slow urbanism',[28] 'slow tourism',[29] and so on.

We share Schneider's desire to couple digital life with the ideas and ideals of the slow living movement. For us, slow computing is a suite of aims and a set of principles designed to transform how, on the one hand, we engage and manage computation in all aspects of our lives (for example, home, work, leisure), and how, on the other, society collectively responds and sets the framework in which computation takes place. As we discuss in detail in Chapter 6, at its heart is an ethics of digital care – of self-care and of care for others. We have chosen to link this ethics of digital care to the notion of slowness, because the wider slow living movement is principally about pushing back against the stresses and pressures that seem to be multiplying across everyday life. It is about seeking an alternative path to the speed and busyness of the modern world that prioritizes a different set of values – enjoyment, patience, individual and collective wellbeing, sovereignty, authenticity, responsibility, and sustainability.

The slow movement is about more than managing time and speed. It is also about experience, quality, and a set of expectations and principles. For example, the slow food movement seeks a more leisurely approach to preparing and eating food, but also desires better ingredients and food stuffs (for example, organic, sustainably farmed), more tasty recipes, and a more enjoyable dining ambience, as well as promoting a healthy living lifestyle and opposing the values and economy of fast food. Slowness is about taking a more measured pace, but it is also about enacting a different kind of society, one that is more reflective and tries to create a better quality of life. In this sense, slow computing is about practising politics with a small 'p' (living your life with respect to some aims and principles), but also in a big 'P' sense (actively and openly promoting and debating those principles and how they should be enacted), with respect to the effects of computation and data on society and economy.

As such, there is a general underlying philosophy to the idea, one that has a built-in ethics of care to oneself, to each other, and to the planet. It's not simply a matter of changing pace, but also changing perspectives about what matters and then trying to enact a more sustainable, enjoyable, and fulfilling lifestyle. In contrast to Facebook's original motto, 'Move fast and break things', slow computing therefore embraces Ruha Benjamin's counter-motto of 'move slower and empower people'.[30] For us, slowness when applied to digital lives is about countering acceleration *and* extraction. It is about creating a more productive relationship with digital technologies; about using devices and apps without feeling harassed, stressed, coerced, or exploited.

The term 'slow computing', then, seeks to capture the diverse actions of a wide range of people who are aiming to moderate, oppose, evade, alter, or otherwise navigate their way around problems such as acceleration and extraction, and other issues encountered when living digital lives. Following our twofold division of the key harms, these resistances take the form of slowing down and temporally reconfiguring digital participation *and* performing intricate 'dances' around and within the proliferating platforms and constituent units and infrastructures of data extraction. Neither of these moves is problem-free, with both practices posing difficulties for technology users. Nevertheless, slow computing is a viable and necessary option if you want to interface with computing on *your* terms.

The axes of acceleration and extraction mean that digital life is coming toward you at a rapid pace and in ways that can seem like you're losing control. How *can* you respond? It is important to recognize that you already have *some* scope – and possibly more if you connect with others effectively – to counter the pressures bearing down on you.[31] In the first place, we will argue that responding does mean taking individual ownership of and responsibility for the issues. You are the one feeling time- and work-stressed from always interfacing with computation; your choices about what data to share are at issue.

That's not to deny that some of this pressure is asserted by others, such as employers, or that some data are extracted whether we like it or not. To be sure, we are bound up in structural conditions and relationships that place limits on our agency and our ability to resist the ways in which digital technologies impact on our lives. For example, governments can compel us to engage with them through digital systems, employers can insist that we are always connected and responsive, and family and friends can pressure us to use particular platforms and services. Different groups also have varying levels of autonomy and capacities to exert control over their digital lives. Those with insecure jobs can be more tightly bound to a digital leash, people of colour and ethnic minorities are more likely to be profiled and targeted in negative ways, and poor and marginalized communities are less likely to possess the tools and skills to practise slow computing. Others possess the social standing and resources to live digital lives that are more under their command.

Nevertheless, we are all active agents in our own lives; we have some ability – even if to a limited degree – to shape relationships, counter prevailing conditions, and make decisions about how to act. There are steps we can all take, many of them minor, that enable us to take back some control and autonomy. Some of these steps will give rise to inconveniences, irritations, hurdles; a few are difficult. But in many cases it's simply about learning new practices, including new 'click-ways' (that is, you need to click here, then there, then choose this option, then that one, and as you click your way through options you start to practise slow computing), or learning when you simply need to stop clicking and just 'switch off smart' (that is, not just switching off digital devices but doing so in a way that lets others know why you're not responding). As you read the following chapters you might recognize some of the slow computing tactics we detail and realize you have already been practising them without necessarily thinking of them in these terms.

While we think individual responses are a crucial element of slow computing, there is no doubt that collective moves are also needed. Some of these are practical; others are institutional, political, and philosophical in nature. They can be led by communities, companies, non-governmental organizations, political parties, governments, or civil society. Some involve new partnerships.

For example, individuals can come together to create open data or develop open source or citizen science interventions that alter how digitally mediated services are delivered. Companies can implement market-led regulation and voluntarily adopt and promote practices such as privacy-by-design (baking individually controlled data extraction into the product) as a means to achieve corporate social responsibility, as well as competitive advantage. Non-governmental organizations and foundations can produce or facilitate privacy enhancement tools such as ad blockers, cookie blockers and removers, malware detection and interception, site blocking, encryption tools, and services to opt out of databases held by data brokers. Political parties can adopt slow computing ideas as policy proposals and push for their adoption by states. Governments can formulate and implement policy interventions relating to fair information practice principles, privacy-by-design, and working hours and conditions, as well as enact new legislation that protects people's rights, and adopt modes of governance in relation to its own programmes and practices that implement and promote slow computing. Academics and civil liberties groups can expose problematic issues and practices, and formulate and debate an ethics of care that sets out the ideal principles for creating and enacting slow computing.

Some communities have been actively practising slow computing since the invention of digital technologies. Their approach demonstrates that slow computing can be actively executed and, depending on how deeply the values are held and enacted, can radically change everyday life. For example, Orthodox Jews practise the non-use of digital technologies on the Sabbath for religious reasons. Interestingly, digital technology is also the solution to the potential downside of observing these beliefs and practices, with automation being used to record television shows and phone messages for later consumption, or for automatically turning lights on or off.[32] The Amish also practise the non-use or limited use of digital technology, partly for religious reasons, but also to preserve cultural autonomy in the digital age.[33] As Howard Rheingold[34] observes, the Amish are 'techno-selectives' rather than technophobes, using digital technologies that they believe will bring the community together and sustain Amish values or that are an economic necessity, and avoiding those that they think will create distraction from family or community time. Important here is not just how technology is used, but a sense of how it may change the person or community. When digital technology is used, such as a mobile phone or laptop, it is often a shared resource and used in a particular, limited way.[35] For example, a single mobile phone might be shared by all workers and only used for work-related calls.

You do not have to use digital technology in the narrow ways adopted by the Amish in order to practise slow computing. However, it does mean being techno-selective in how you use technologies. Moreover, there are lots of moves underway that are not bound up in religious beliefs and cultural communities but actively enable slow computing practices, and most of them will appreciate your support. This support can vary and can include: active participation in creating interventions; helping to fund open source projects; promoting the agenda or principles of slow computing; or adopting new tools and practices in your everyday life. Other kinds of intervention will no doubt emerge as digital life motors on. Trying to build toward slow computing means staying alert to these developments and taking steps with others to lay the right foundations and then to strengthen them where or when they are needed. The onus really is on all of us to find ways of responding together in imaginative ways, while we still can. A key point to make here is that the joy of computing – the speed of access; the spreading of ideas; the sharing of information and insights; the way we can connect with others – creates serious scope for all of us to bounce ideas around and debate and develop new practices and principles. It might be tempting to keep clicking or swiping without any thought for what's taking shape around us, but our hope is that fewer of us will take that approach. Only via individual and collective actions will we be able to shift gear and create a more balanced digital society.

To be clear, then, we are not advocating for total withdrawal or making only slow progress. As Carl Honoré, a slow movement guru, argues, 'Slow is about relearning the lost art of shifting gears. … Speed is wonderful, thrilling, liberating, fun, and it can be immensely productive.' Like him, we are not anti-speed. We agree that, 'you've got to have a range of speeds. Like any piece of music, you can't have just one tempo.' For Honoré, slowness is about 'doing things at the right speed – what musicians call the *tempo giusto*' and avoiding the 'trap of trying to do more and more things in less and less time, putting quantity before quality in everything we do'.[36] What we're advocating for is a new way of moving and ultimately more effort to become reflective about what the journey through digital life entails. It is about recognizing that slow computing can obtain real benefits for individual and societal wellbeing.

Overview of the book

So what actions are possible, what might we *need* to start doing together, and what obstacles are in the way? We answer these questions in the following chapters. We begin with the problems and obstacles.

Chapter 2 explores the acceleration of everyday life. We highlight what's been happening over the last few years and explain why our lives do seem to be speeding up, examining the transformation in the pace, tempo, scheduling, and connections of work, home, and social encounters. In turn, we explore the positive impacts of these changes, such as quicker movement, on-the-fly encounters, more efficient and timely accomplishment of tasks, and the optimization of services, before examining some of their negative outcomes, such as never-ending engagement, time-stress and scarcity, little time for contemplation and deliberation, and the introduction of more technocratic forms of governance. We note that there is probably no way of *entirely* getting around acceleration while also making use of – and finding some joy in – digital life. But striking a balance? That's within reach.

Our focus turns to the theme of data extraction and its significance for slow computing in Chapter 3. We examine how digital technologies are producing digital footprints and shadows about our everyday lives and how these data are put to use within multibillion dollar data brokerage and advertising industries. We explore the apparent benefits of trading data for services and also its dark side with respect to four key issues: privacy; difference, profiling, and sorting; governance and politics; and production. These issues present significant obstacles for slow computing, but also gateways through which a more balanced digital society can take shape.

In Chapter 4 we identify a number of specific moves that all of us, mostly on our own as individuals, can make to push toward slow computing. Our attention is first on the temporal component: on the slowing down of things. We introduce the idea of 'time sovereignty', and emphasize reflection, auditing, and identifying contingencies that provide more control over the pace and tempo of digital life. Next, we focus on data extraction and how to manage or evade excessive data harvesting, utilizing the concept of 'data sovereignty'. We outline four sets of interventions that we can all perform to try and protect ourselves: curation, using open source alternatives, stepping away from digital technologies, and practising obfuscation. Even using these tactics, slow computing can be hard, and we examine why this is the case, before finally considering how you might go about contemplating and formulating your own slow computing strategy and the tactics required to realize it.

Moving from the personal to the collective, Chapter 5 outlines a wide range of practices you can participate in as you move toward a more balanced digital life. Some of these practices are for the expert technology user, but most of these need support from the casual user. One trope about digital life is our apparent powerlessness against the might of corporate giants or governments. However, we're already seeing moves by lawmakers in some countries to alter the terms on which citizens are expected to lead digital lives. Collectively we can make a difference. We examine two connected components of collective moves to decelerate: slow computing practices and rights, and the creation of slow computing spaces. We then consider collective actions to evade data extraction from different perspectives: industry-led moves; regulation by government and policy makers; and data sovereignty expressed by communities, civil society, and non-governmental bodies. Finally, we consider what a slow computing world might look like.

Thinking about the future of digital life is the focus of Chapter 6, where we develop a normative argument about what our digital society and economy *should* be like. Here, we try to reimagine digital life through an ethical lens focused on care, fairness, equity, and justice. Our aim is to provide the moral arguments for creating and sustaining slow computing. We first set out an ethics of time sovereignty that justifies slowing down, then an ethics of data sovereignty that provides users with more autonomy and control over their data and its uses. We explore the extent to which slow computing can extend to all, regardless of gender, race, class, or abilities. Do we all have the same opportunities to tackle acceleration and data extraction?

In the final chapter, we conclude our case. Having identified a range of personal moves and collective practices, and after emphasizing the need for a new ethics of digital care to take shape, we explore the creation of a more balanced digital society and some of the persistent obstacles to implementing slow computing and how to overcome them. The joy of slow computing is within our grasp. It is something for all of us to work toward. But to realize it we need to understand the issues and react to counter the forces of acceleration and data extraction. Hopefully you will join us on that journey.

2

Accelerating Life

Tuesday. 7.45am. Your commute is going well. No delays so far. Of the 15 emails awaiting your attention, you've answered eight. Only seven more to go. But these are tricky ones. You dip into your phone's browser, download a file from your work's server, scan it to ensure it has the data you need. Back into email. 'Hi Sarah, sales were up 3.4% in March. Let me know if you need further info.' Back into the browser. Apparently, there was an angry message posted on Twitter last night. You find it, look through the profile of the customer, copy the message. Back into email. 'Hi James, some flak on Twitter last night. I'm pasting the message. Please resolve.' You see that Sarah has replied already. She wants to know what sales were in March for the last five years. You'll be arriving soon. No time for this. 'Sarah, I'll come back to you in twenty: just arriving on train.' James has replied also. 'Who wrote this tweet?' Back into the browser. No internet connection? You look up: other commuters are looking up and around, confused. Seems the train's Wi-Fi has gone down. Only then do you notice: you missed your stop.

For many people today, the separation of time spent at work and time spent away from work has disappeared. Before the digital revolution, a fair few workers would have worked during their commute: reading a report perhaps, or preparing notes in advance of an afternoon meeting. But there were technological limits to this type of activity. With no email, your connection to a Sarah or a James hinged on meeting them once you arrived in the office or giving them a call from your desk. While reading a report, you weren't receiving 'pings' asking you to handle some other matter. Nowadays, technology facilitates and encourages multiple simultaneous conversations across numerous platforms. Once you're online – which could be from the moment you wake up – work activities can easily suck you in: answering emails, checking technical reports, looking for updates, requesting information from colleagues, editing files, and more, all while you make coffee, eat breakfast, get the kids dressed, take them to school, and head for the train. Alternatively, if you are working on zero hours contracts or on shift work, the days and hours you work could be communicated to you at short notice, meaning you have to quickly rearrange your day with all the stresses that go with that.

The sort of 'working time drift'[1] or working time flexibility we're describing here is a growing feature of social life today. In the UK, it is estimated that workers undertake on average 5.1 additional hours a week of work beyond what they are contracted to do.[2] In some sectors it is notably higher than this. Some of this work might be paid as overtime, in other cases it is non-remunerated. In many cases, digitally mediated jobs are not counted, given they consist of numerous small tasks – two minutes here, five minutes there – rather than large blocks of time. The digitally mediated nature of work can pose a serious challenge for many of us as we try to strike an appropriate 'work-life balance'. The risk is that we engage in 'limitless work',[3] when labour encroaches on and consumes our early mornings, evenings, and weekends. This is already the case for some workers. It also seems to be getting worse, with employees in more and more sectors of the economy drawn into workflow systems that expect – or at least invite – us to be 'always-on'[4] and be 'everywhere available'.[5] A simple commute, like the one mentioned in the opening epigraph, can now require rapid and intense activity.

This working time drift has become culturally expected, promoted through a set of values that valorizes overwork as something to be admired and celebrated.[6] The way one gets ahead in life is to put in Herculean work hours, be always available, and exploit one's own time and labour. Long hours, failing to take vacations, stress, and 'overwork' – indeed being a workaholic – are thus framed as badges of honour. If one burns out, then that is the price of striving for success. As slow advocate Alex Soojung-Kim Pang notes, this view is reinforced through a sense that 'a perpetually uncertain economy requires us to accept these terms or be replaced'.[7] As Arianna Huffington notes, 'A global economy that never rests expects us to ignore the need for rest'[8] and to continually create value through work and other activities. Moreover, the digital leash and a plethora of apps now enable leisure time, night-time, and even sleep to be colonized and monetized by digital enterprise. This is reflected in the shrinkage of sleep time, with Americans today sleeping on average 6.5 hours per night, compared with 8 hours a generation ago.[9] Part of the issue is that we are not just rewarded for performing work, but for performing busyness, which is measured through the devices we use (tracking key performance indicators such as key-strokes, logins/-outs, volume of trades, and so on) and the hours we drift over our contract. As Pang observes: 'In the modern office, all the world's a stage, nowhere is off camera, and the performance never stops.'[10] Other aspects driving overwork are decreased job security, increased competitiveness, demands for productivity, flat wages, and rising prices for housing and commodities, along with the erosion of labour unions, all of which force us to work harder to retain employment.[11]

It's not just work that is the source of shifts in how time is experienced. How we interact with each other and organize our social life, aided by new digital technologies, is also leading to a change in scheduling and planning, and the pace and tempo of life. Mobile phones and locative media, such as location-aware smartphone apps, have altered the practices of coordination, communication, and social gathering in places by enabling on-the-fly scheduling of meetings and serendipitous encounters by revealing the location of nearby friends.[12] And digital technologies enable us to timeshift activities to formerly unavailable time slots. For example, we can shop online or pay a utility bill while waiting at a bus stop. What was 'dead time' becomes 'productive time'.[13] Moreover, these technologies empower us to multitask and interleave activities so we can perform several tasks simultaneously rather than sequentially: now we can be watching television while also posting on social media, looking up potential travel arrangements, snacking on junk food, and chatting with family. In addition, the networked monitoring of city systems using sensors, actuators, and cameras means that we are able to use apps to access real-time information about present conditions – the location of buses/trains, how many spaces are in a car park, how many bikes are in bikeshare stands, what the weather is – and can then react accordingly.[14] We no longer need to plan routes or activities; we can make decisions in the here and now, based on what is currently happening. While there are benefits to these changes, there are also negative consequences, a notable and perhaps ironic one being that these technologies often create time stresses – a feeling of being harried and rushed – rather than more leisure time.[15]

Against this backdrop, our aim in this chapter is to provide an account of how and why social life has been speeding up and, in turn, why it needs to be slowed back down. Academics have been studying these temporal trends and they've been offering a wide range of explanations as to what is occurring. We cut into these theories and draw out an analysis of the core forces playing out in front of us today. Our point is that acceleration undoubtedly impacts on the types of lives we can lead.

We start the chapter by highlighting the technological developments giving rise to acceleration. There are two sets of developments to note. The first refers to the path-breaking transportation and communications technologies that have shrunk the world so significantly over the past two centuries. Railways, cars, planes, telegraph, and phones have all played a huge role in creating what is termed 'time-space compression'[16] in which the world has become smaller and faster. We'll refer to this set of technologies as 'analogue accelerators'. The second set are 'digital accelerators': they use today's computing power to amplify and even transform how we move and connect with others. They include the internet and personal computers, mobile phones and smartphones, and (soon) autonomous vehicles. As well as increasing the speed of communication and transfer of information, these technologies are mediating and augmenting analogue modes of travel, logistics, and service provision, enabling online purchasing, quicker access to information, and more efficient and optimized delivery and coordination. Services can be accessed at a distance, saving time and effort, and also bypassing physical queues.[17]

As we then discuss, when we focus on acceleration we find positives and negatives to consider. One aspect of the joy of computing, after all, is undoubtedly that technologies can empower all of us. Speed is one key component here. Put simply, acceleration does enable us to accomplish more tasks. There is a lot to like about that. But there are costs, too. Accelerating technologies increase the demand for responses. That can be exhausting. Many people can find the source of a growing sense of anxiety arising precisely from acceleration. In short, there is a constellation of effects to ponder. We use these points to then argue that creating a balanced digital life requires making new and ongoing calculations about which effects require the most attention in your life.

Faster, smaller, interconnected

Acceleration stems from invention and from innovations and their (often rapid) adoption. This relates in particular to workplace technologies designed to increase productivity (such as the assembly line), and workplace time regimes designed to manage workflow (such as clocking in and out, and workload models). Businesses that speed up operations, optimizing time and creating efficiencies, gain competitive advantage and profit. Time is money, as the saying has it. As such, companies have invested in creating and adopting technologies and practices that speed up the process of producing goods and services and getting them to market. Indeed, acceleration seems imperative in order to survive and thrive within capitalism. First steam engines, then combustion engines and electrical devices, transformed the operations of factories and offices, and also the productivity of workers. Digital technologies are the latest key means of driving workplace acceleration: they can undertake calculations far more quickly and efficiently than people, can be used to control and automate many other systems, can foster innovation enabling companies to stay ahead of competition and obsolescence, and can effectively monitor and manage workforce performance. No surprise that US corporate investment in IT increased from $5 billion in 1970 to $350 billion in 2008.[18] By 2011, 76% of full-time workers and 52% of part-time workers in the US used the internet at work, and only a quarter of all workers used no computing device at work.[19]

Home life has also been transformed by a series of technologies designed to speed up and reduce the drudgery of domestic tasks and create more leisure time and improve quality of life. Manual cleaning is largely replaced by vacuum cleaners, washing machines, and dishwashers. Preparing food and cooking are performed using fridges, freezers, blenders, electric stoves/kettles, and microwaves. Bathing has switched from heating water over fires for baths to on-demand hot showers. Outdoors, we tend our gardens using (nowadays, even automated) electric mowers and cutters and sprinklers, and clean driveways using leaf blowers and power washers. And our quality of domestic life has been improved through more responsive and controllable heating and air conditioning systems and a plethora of in-home entertainment media (streaming radio, television, internet). We no longer need to gather fuel or travel for leisure. And we are much more easily able to multitask and interleave tasks; for example, unstacking the dishwasher and loading the washing machine in between cooking tasks, while talking on the phone and catching up with the news on the television in the background. Technologies have also reconfigured domestic dwelling arrangements, such as skyscraper apartment living enabled by elevators, and suburban housing enabled by public transport and cars. Domestic technologies are becoming digitally augmented so that they can be programmed and used in more diverse and sophisticated ways. They are also becoming connected, enabling us to control them from a distance (such as turning on a coffee machine or changing heating settings from the bus on the way home).

Another essential driver of acceleration has been innovations in transportation and communication technologies designed to speed up the circulation of goods, bring people closer together, hasten the flow of information, and interlink places. Over the past two centuries, a series of major technological innovations have created significant time-space compression: that is, a shrinkage in the time taken to traverse and connect across space, making the world seem smaller and more interconnected. Compared to the world our grandparents and great-grandparents encountered, it is much easier, quicker, and cheaper for us to move and communicate between any two points on the planet. Finding information and completing tasks such as shopping can be undertaken without leaving home. As with accelerating technologies used within offices and factories, time-space compression is critical to capitalism and economic growth.[20] Movement and transportation time costs money, so accelerating the speed of travel and communication saves expense while also opening up new markets, thereby producing profit. Acceleration is driven by the economic logic of capitalism.

Scholars note that time-space compression consists of two related processes. First, time-space convergence is the shrinkage in time taken to communicate or travel between locations.[21] Since the development of canals and railways the mass movement of people and goods has got progressively quicker and cheaper, and since the invention of the telegraph, communication between distant places is becoming real-time in nature. Second, time-space distanciation is an increasing synchronicity between places so that they become interpenetrated and interdependent.[22] For example, companies have been able to organize their operations across the globe, with workers in one location overseen by managers in another, and to centrally manage vast, complex logistics networks. People, goods, services, information, and money flow between locations creating interdependencies, so that what happens in one place has a direct effect on what is happening elsewhere. In essence, these two processes have meant that the friction of distance is progressively overcome through speed: or as some would put it, space is annihilated by time.[23]

Up until the 1970s, time-space compression was driven by innovations in analogue technologies – canals, railways, telegraph, telephone, cars, planes, radio, television, satellites. Canals and railways were a key driver of the Industrial Revolution and opened up the interiors of continents to trade, resource extraction, and colonization. Until their widescale rollout, the fastest way of moving across land from one place to another relied principally on horses. Travelling speeds by horse and carriage, along good roads in dry weather, might have been around 10km per hour.[24] Distances across space were, accordingly, quite an obstacle to the rapid movement of people, goods, or ideas. In 1750 it would take 48 hours by stagecoach to get from London to Bristol, a distance of 118 miles. By 1821 it took 24 hours due to improvements in the roads.[25] Good quality roads only connected key places, however, with travel slower off the beaten track or over hilly countryside. The fastest way across water prior to steam power was by sailing ship, which in turn depended on prevalent wind, weather, and currents. It could take two to four months to cross the Atlantic. Long-distance travel was exceptional, costly, and often in a single direction: those migrating to the 'new world' from Europe knew that the trip was likely to be one way and they would never see their homeland again. They also knew relatively little about where they were travelling to, given the paucity in the flow of information; any communication back home would be via letters that could take weeks or months to be delivered.

A truly transformative shift was the invention of rail travel. By the mid-1800s, it took five hours to travel by train from London to Bristol, and by the 1880s just two and a half hours. In 1830 it took three weeks to travel from New York to Chicago by stagecoach; by 1857 it took two days by train.[26] The steam-powered railways annihilated space with speed, and did so for large numbers of people on a fairly regular basis. Today, in those areas of the planet where travellers are lucky enough to find them, high-speed trains travel at over 300km per hour. Sailing times tumbled with the introduction of steam-powered ships. Whereas a trip from England to Australia could take several months under sail, by 1882 it was 30–40 days to Perth and 40–50 days to Sydney.[27] By container ship, the journey can now be done in less than three weeks via the Suez Canal.

Soon after the turn of the 20th century, the combustion engine, the motoring heart of the automobile, revolutionized personal travel: the car gave travellers the capacity to travel on their own or in small groups and scope to explore new spaces that had not been penetrated by the railways. Over time, the road network became more extensive, vehicles became less expensive and more reliable, and along key routes such as motorways speeds averaged up to 100km per hour, meaning they moved across land at least ten times faster than horses, and could do so for hour after hour. The road network also facilitated door-to-door delivery of goods.

By the 1920s and 1930s, a new network of airports and routes started to span continents to facilitate early passenger air travel. While quicker than sea or rail travel, in 1934 it would still take two to three days to hop across the United States, and eight days to travel from London to Singapore.[28] And it was by no means a cheap form of travel, with a ticket on the latter route costing £180 (which was about 1.4 times the average yearly British wage in 1937).[29] The 1930s also saw the first airmail services, further reducing the time to deliver letters and parcels. By the 1960s, jet-propelled aircraft were crossing oceans and continents at almost 1,000km per hour and it was possible to reach Australia from Britain in less than a day. Although the world's poorest people are still unlikely to afford such an experience, almost everyone on the planet has seen such aircraft in the sky and a couple of billion people have travelled at that speed.

The quickest form of communication prior to the telegraph was the use of carrier pigeon, though this had a natural range and was typically unidirectional (from where you were to home); otherwise it was postal mail via stagecoach. Invented in 1837, the telegraph enabled messages to be sent along an infrastructure of wires, being passed between operators along a chain. It drastically reduced the time needed to send information from one place to another (as long as each place was on the network). For example, in 1870 it would take 50 days to send a letter from London to Hong Kong, but just three days to send a message via telegraph.[30] Over the course of the 19th century, telegraph cables were installed between every major town and city, with cables laid across ocean beds to link continents. They became a vital infrastructure of government and business due to their timely communication. The telephone, invented in 1876, transformed messages from Morse code to voice, enabling verbal communication with someone not immediately in earshot and who might even have been on the other side of the world. Throughout the 20th century, the penetration of the telephone slowly increased as new lines were installed, making its way first into wealthy, then middle-class and finally working-class homes.

Radio, invented in the late 19th century, was a key broadcast medium. In 1894, the first commercially successful wireless telegraphy system was produced by Marconi, with applications for military and maritime use. In 1920, the first radio news, sports, and live concert programmes were broadcast to the public on systems that reached up to 100 miles. By the mid-1920s, radio stations were starting to be linked together into networks, with programmes being simultaneously broadcast across the relays, and by 1931 networks in the United States had become coast-to-coast in extent (though much of the country, especially in the west, was off the network).[31] By 1940, 80% of households in the US owned a radio, whereas only 36% owned a telephone.[32] Radio was timelier and more far-reaching than the circulation of newspapers.

Television also has its roots in a series of inventions in the 19th century, but it was not successfully transmitted as a fully working system until the mid-1920s. Its initial commercial development was slow, with the first publicly available black-and-white television sets and broadcast stations appearing in the 1940s. Colour broadcasts were begun in the mid-1950s, but only started to become regular when the first reasonably priced colour television sets were made available in the mid-1960s.[33] Over time, innovations such as cable and satellite televisions and video recorders widened viewer choice.

All of these technologies changed how people understood and experienced distance by shrinking and speeding up the world. By the mid-20th century a new sense of the globe had emerged, with people, goods, information, and money crossing land and oceans along numerous new and expanding pathways. People could travel faster and further, they had access to more information, goods, and services, and their lives were ever more interconnected with other places. As networks expanded and the volumes of people using them increased, costs plummeted, making them accessible to more and more individuals. Today, we can travel halfway round the world for a few hundred euros or dollars, and can communicate vast amounts of data for a few cents.

And yet, if increases in the speed at which humans can travel and send information in analogue forms have been phenomenal, they seem mediocre when compared to the volume and speed at which we move information today, or coordinate and optimize travel. The development of computing technologies unleashed digital accelerators that rapidly overcame the limits of analogue acceleration. Since the 1970s, digital accelerators have had an increasing influence on everyday life.

The internet was invented in the late 1960s, with the first connection between computers in Stanford and UCLA in November 1969. By January 1971, there were 13 nodes in the network and by June 1974 there were 62 across the United States, with at least two available routes between all nodes.[34] The TCP/IP protocol that still enables a vast diversity of machines to communicate and transfer data with each other was developed in 1973.[35] Email was made operational across the network in 1970. Other significant technological developments during the 1970s included: the development of Ethernet, allowing the first Local Area Networks (LANs) to come online; the sending of data via satellites; networks being established outside the US in France, Germany, Japan, and the

UK; and the first non-institutional networks being developed by hobbyists using personal computers, modems, and telephone lines to make early bulletin board systems (BBSs).[36] BBSs were the forerunners of general file sharing, public-access services, and social media. By the end of the 1970s, the internet was still largely closed, with access limited to research and military personnel and hobbyists. The 1980s were characterized by a steady growth in both institutional networks and BBSs, the development of public-access internet architecture, and the construction of nascent virtual gaming in textual worlds.

The main breakthrough, in terms of generating the exponential growth in internet content, services, and usage, came in 1992 when Tim Berners-Lee at CERN released the protocols and software that created the 'world wide web' (WWW, or the web).[37] Here, text, images, and sound could combine to provide a range of information, with hypertext used to link sites together. Mosaic, a browser application providing users with an easy-to-use, graphical interface for webpages quickly followed, as did search engines. The first online shopping started to appear, with Amazon launching in July 1995. In March 1995, the web became the service with the greatest traffic on the internet, overtaking file transfer.[38] In the main, the web was still a broadcast medium, with companies and individuals creating webpages that others could browse and explore, and participation was relatively limited. Around 2003, the nature of the web started to change as it transitioned to what has been called Web 2.0.[39] Rather than simply browse information or practise consumption, parts of the web became a read+write media, in which people added value to sites as they used them (through, for example, social networking sites such as Facebook and Twitter) and could more easily produce their own or shared content (photo and video sharing sites such as Flickr, Instagram, and YouTube), wikis (open, collectively authored websites, such as Wikipedia or OpenStreetMap), and blogs.[40] With the advent of the smartphone in the 2000s, the internet has become mobile, accessible from anywhere across the planet through wireless connectivity (such as GSM [Global System for Mobile communications], Wi-Fi, Bluetooth) and a diverse range of apps. By 2016, just over two billion people owned a smartphone,[41] using them to undertake many different tasks that extend well beyond making phone calls.

Beyond people's access to and interaction with the web, objects, devices, and infrastructure are now becoming part of the 'internet of things'. In 2018, there were already estimated to be 23 billion networked objects making up the internet of things, with this figure set to rise to 50 billion by 2023.[42] Each 'thing' connected to the internet is assigned an exclusive identification code and can be communicated with remotely.[43] What this means is that its use can be monitored and controlled from a distance, turning what were previously quite basic consumer items and household goods like fridges and heating systems into 'smart' objects, and enabling formerly analogue systems to be optimized through computation. In turn, systems that used to be controlled exclusively by human oversight start to become automated, with data streaming from sensors, actuators, cameras, and devices to be processed by algorithms, and instructions flowing back.[44] For example, in an intelligent transport system, data flows in real-time from inductive loops in the road, traffic cameras, and transponder boxes to a control room where specialist software monitors the flow of vehicles and calculates how to best optimize traffic signalling, sending phasing instructions back to the network of traffic lights. In this way, the system tries to automatically adjust to current traffic conditions. Such developments are making real the notion of smart cities and smart homes.

More recently, there has been significant investment in the creation of autonomous vehicles and drones. Networked computation and spatial big data (such as GPS, LiDAR, high resolution maps) are enabling driverless movement. To a large extent this remains a sci-fi trope rather than a reality and there are significant regulatory issues to be addressed; however, a number of testbed trials are already underway, with companies investing in the possibility that such technologies will transform personal mobility, logistics, and consumption. What this means is that, more and more, we live in a world of pervasive and ubiquitous computing, where computation is embedded into things and available everywhere.

For some, the time-space compression enabled by digital accelerators has been so profound that they suggest that time, not space, has become *the* crucial dimension and resource of social and economic life.[45] The internet and rapid travel mean that we live in an age of 'instantaneous time', which holds different qualities to 'clock time',[46] for example: synchronicity across time zones (people across the globe can experience shared moments, such as simultaneously watching a sporting event or media story, or interacting via social media); a breakdown of distinctions between day and night, weekdays and weekends, and flexibility in working hours as employment and social practices change; volatility, disposability, and mobility of fashions, products, and ideas; an erosion of established temporal norms such as family meal times; just-in-time logistics and either instantaneous delivery of goods and services (via the internet, such as digital music, books, TV/movies) or speedy delivery (within hours or next day); and real-time management of infrastructures and systems.

For time theorist Robert Hassan, just as the clock shifted our relationship with time from social (such as festivals, holy days) and natural registers (such as Earth seasons, diurnal cycles, body clocks) to an abstract mechanical register ('clock time'), networked computation is shifting the temporal logic for society and economy from 'clock time' to 'network time'.[47] Set meal times, clocking in/out, timetables, pre-arranged meetings, and so on, built around the measure of a clock, are being traded for greater temporal flexibility and timeshifting (events being organized and coordinated on-the-fly across space and scales). Network time shifts the scheduling and planning of activities and events from 'specific times and places' to 'any time, any place'. And it means that activities such as shopping, communicating, banking, play, and travel are now operating in a distributed 'perpetual present'[48] – instant, always, and everywhere.[49] In turn, our sense of and experience of time are changing. As such, digital accelerators are having a huge impact on our world.

The positives of acceleration

Analogue accelerators went a long way toward shrinking and speeding up the world. Digital accelerators have taken things much further. We are not alone in finding a lot to like about these facets of acceleration – companies are investing in, and governments are supporting and adopting these technologies because they know there are significant advantages to be gained from their use and there is strong public appetite for them. You probably feel the same. Boiling things down, at a personal level there are at least three major positives of acceleration, relating to moving, connecting, and accomplishing.

Moving: From the railways and cars to airplanes, acceleration has enabled humans to move across land and ocean at fantastic speeds. There is obviously a lot to like about this. You can live 100 kilometres or more from your place of work and commute daily. Or you can work at home (or in a coffee shop, or hotel room, or on a plane, or indeed anywhere with a mobile phone signal or Wi-Fi) and yet be in permanent connection with your workplace. You can work simultaneously on the same document with colleagues who are located on five different continents, or take part in a conference call. You can even work simultaneously on that document while you are on the move, commuting to or from work (or stream music or video or browse the internet). Your sibling who moved abroad is certainly still far away but, so long as they can afford it, they can get back quickly in a family crisis. You can easily visit your child who is attending university in another city.

Freed from clock time you no longer need to organize meeting a friend at a certain location and time in advance, but can call or message them while you are in transit and arrange meeting up. While travelling by public transport you do not need to buy a ticket, you just tap-in, tap-out with a smartcard that keeps track of payment, and you can access real-time information about when the next train or bus is, or get suggestions of the best route to take. If you are travelling by car, you can find out where the congestion is and take an alternative route. Once you have met your friend you can use a location-based services app to get recommendations for nearby restaurants. If you are visiting a place by yourself you can use your phone to see if any of your friends are nearby and then make contact. In terms of tourism, we no longer need to visit a travel agent: we can explore thousands of options from the comfort of our beds, as well as compare multiple prices from different sellers, and book instantly. Some goods might not be available in your local stores, but you can easily go online, track them down, and order them and they will be delivered within a couple of days at most. In many cases, you can even customize them, choosing specification, design, and colours. Many of these tasks were impossible in the pre-digital age, and others were certainly more constrained. Movement has never been so easy, fast, flexible, responsive, adaptive, and cheap.

Connecting: New features of social life become possible when we have access to high-speed networks. Video communication with loved ones on the other side of the world, often in vivid colour and crystal-clear sound, is a regular feature of life for millions of people today. We can use the internet to track down old friends, connect via social media, and keep in contact with them, sharing memories and letting them know what we're doing or thinking. If we are fans of particular bands or movie stars, we can follow their social media accounts and even get to chat to them directly, or engage in a debate with a politician who we'd rarely see otherwise. We can use the internet to research our family trees and locate long-lost cousins, great uncles or aunts and make contact and arrange to meet them.

Email, texting, social media, and messaging channels can be used to create a continually evolving conversation among friends and strangers. We can share experiences with thousands of people scattered across the globe by simultaneously watching an event or television show and discussing it via social media. Or we can become our own broadcaster, sharing our thoughts, opinions, and creative outputs (music, video, art) via blogs, social media (such as Facebook, Instagram, Twitter), YouTube channels, and podcasts. Some people have even managed to turn their broadcasts into a career, using their content and celebrity (as evidenced by the number of followers or subscribers) to leverage advertising revenue. Children are now expressing a desire to become a 'YouTuber' or 'social influencer' when they grow up. We can leave reviews of products on retail sites, or specialist review sites such as Goodreads (books) or IMDB (movies), or write comments on blogs or boards and engage in conversations and debates.

In more social and political terms, we can use various mobile phone and internet media to organize civil protest and activism. The Arab Spring is perhaps the best-known example, a revolt consisting of millions of citizens across several countries in North Africa and the Middle East, orchestrated on and through social media.[50] At the time of writing, citizens in Hong Kong are using various forms of encrypted social media to coordinate protests.[51] If we are looking for finance for a good cause or to start a business we can seek to raise funds through a crowdfunding drive. However we use the internet, digital accelerators make it possible to connect with others in ways that were unattainable in the analogue age, all for the cost of our devices and a data plan.

Accomplishing: Maybe you've moved home. Or perhaps you're planning a party. Or you're trying to find the address and opening hours of a shop, or the nearest pharmacy or library. Whatever the reason, you will probably have had similar experiences to us: going online, checking for information, searching for the best option and deal, booking services, ordering deliveries, and so on. Or perhaps you are after a particular tool that will help complete a creative task, such as a photo, video, or sound editor. Or you're looking for an interesting new game to play, either by yourself or with others. Maybe you have a broken machine or car and need to find a manual or advice on how to fix it, or you are after a recipe for a particular kind of cake. The internet and the digital accelerators arranged around it make accomplishing things (in theory, if sometimes not in practice) quicker and easier. The same applies to work-related tasks. Emails or Slack messages ping back and forth as tasks are managed and completed. Skype calls help resolve problems. Databases located in the cloud are tapped into; information is extracted quickly, processed, and converted into value. Specialist software can be discovered and used. Workers can use handheld devices or phones to take photos, record sound, enter information, communicate with colleagues, work on files, and so on. Digital acceleration means we can get many tasks done in ways undreamt of until recently: more speedily, efficiently, and remotely.

Moving, connecting, and accomplishing using networked computing in all these ways enable us to be enmeshed in several competing temporalities simultaneously. For example, you might be heading to a meeting at 10am, using your mobile phone to talk to a colleague on the other side of the planet, while waiting at a pedestrian crossing for the network-controlled traffic lights to change. Here, you are negotiating global time and local time, clock time and network time, as well as social and natural time. You are experiencing the pronounced time-space distanciation of a long-distance call, as well as the very localized time-space choreography of negotiating a traffic junction. Over the past couple of decades, we have become used to negotiating these kinds of complex 'chronotopia' of varying pace, tempos, rhythms, scheduling, temporal relations, and modalities.[52] And while some of the tasks we've discussed can be exhilarating, in the main, they have also become mundane and routine. They are just a part of everyday life. Nevertheless, they are aspects of life today that you, like us, most probably value and appreciate.

Beyond personal fulfilment, accelerating technologies and the time-space compression they create are having profound positive effects on society and economy across every domain of everyday life: family and community relations, economic development, labour, government, education, transport, health, and so on. Networked computation enables more timely reaction, remote access, the optimization of services, new and better products and experiences, and efficiency and cost savings. Rather than explore the benefits across sectors, here we focus on just one by way of illustration. We have selected it because of the profound changes that it has made to Ireland, where we live.

In 1987, Ireland was a relatively poor country on the periphery of Europe. Its GDP was two thirds of the EU average, with only Portugal having a lower rate.[53] The 1980s was a tumultuous decade. Unemployment was nearly 20% at times. There were several changes in government and there was a high emigration rate as younger people in particular sought a better life elsewhere. By the early 2000s, Ireland had the second highest GDP in the EU, its economy was booming, it had almost full employment, and it was experiencing strong immigration of skilled labour.[54] What had created such a transformation in fortunes? In large part, the country had plugged itself into what was then emerging as the 'information superhighway',[55] investing in improved ICT infrastructure and using networked services industries, alongside changes to other economic levers such as taxation rates, planning laws, and special development zones, to drive rapid economic growth. Initially, the country sought to attract low-skill services and high-skill manufacturing to replace an ailing branch-plant economy.[56] It then used this platform to attract higher-skill service jobs and the European headquarters of global tech companies, creating centralized hubs of ICT-led economic activity.

Many major IT companies now employ large numbers in Irish cities, including Apple, Microsoft, IBM, Dell, Google, Facebook, LinkedIn, Softbank, and Huawei, as well as specific IT sectors such as healthcare technologies, and other industries that are highly reliant on ICT, such as banking and financial services. More recently, Irish cities have focused on becoming smart cities, in part to drive economic recovery after the 2008 financial crisis, with local government sponsoring 'hackathons' and procurement-by-challenge initiatives designed to create new start-ups, as well as enabling testbed urbanism that makes city spaces available to companies to trial and test new products as a way of attracting new investment.[57] In effect, the country has been using the time-space convergence and distanciation of ICT to overcome its peripheral location with respect to Europe and the friction of distance by plugging into the global informational economy. In other words, fast computing has had a profound effect on the country, creating wealth and improving quality of life.[58]

In many ways, then, acceleration brings with it major advantages, serious benefits, and new features of everyday life that few of us would want to abandon. No wonder that so many businesses and governments are active in this space: encouraging investment, facilitating innovation, and bouncing off and leveraging each new development in unexpected ways. The fact is, people enjoy the moving, connecting, and accomplishing that digital accelerators make possible. As much as we call for slow computing and a balanced digital life, we're not suggesting we get rid of the digital devices we have available at our fingertips.

The downsides of acceleration

For all that we like what digital accelerators bring us, they also have costs and downsides. There are numerous ways of proceeding here, but we'll begin by putting it like this: digital acceleration, in particular, demands that we all *respond* to the working of technologies, often at an expanding rate. We feel this pressure to respond, even when we're commuting to work or trying to relax at home on a Sunday. And it can be exhausting.

While digital accelerators enable us to accomplish tasks, they also demand action. The more we engage in systems or processes that provide us with rapid responses, the more we expect others – whether humans or machines – to chip in and respond to us. The time we wait for responses from machines, in particular, has decreased significantly. Where once, for example, we had to post a form to a government department to renew our driving licence and then wait for a reply, now we can now do this online and receive an email notification almost immediately. We buy flights, order books, clothes, or consumer electronics online, and as soon as we agree to the purchase we look for a confirmation email. The systems enabling us to accomplish tasks are designed to ensure this rapid response occurs. The danger here, from the perspective of online retailers, is that we'll experience a sense of buyer's remorse, reconsider, and cancel our order. Best to get that notification sent out quickly.

But once we become accustomed to this sort of response time from machines, it becomes tempting to expect a similar response from humans. We email a colleague or send a WhatsApp message to a sibling, wait for a response, and want one quickly. One of the smartest innovations of the last few years – and a profitable one for those who designed it – was the double-tick used by WhatsApp to show a message was not only sent but also received and read by the recipient. When we know that the sender knows that we've seen their message, we feel pressured to respond. Extend that feeling of pressure across the wide array of apps or communications platforms we use on a daily basis and the pressure builds, and can become a major source of anxiety.

Almost anyone using today's devices will know this feeling: numerous prompts per day across the multiple platforms that constitute our digitally mediated lives (email accounts, social media presence, specialist workflow systems, SMS messages, missed calls, bills to be paid online, and so on) that create an endless flow of demands. In our daily practices of using digital technologies we soon become aware that other humans or machines are either standing in the way of us accomplishing tasks or are themselves waiting for us to respond so their current list of tasks can be completed. Acceleration draws all of us into this response loop and creates temporal densification and fragmentation: that is, intense, overlapping temporal demands and rhythms, which force us to deal with multiple, different tasks in quick succession or simultaneously. Networked computing produces ever more extended and complex sets of tasks to attend to and a temporal regime that compels us into a never-ending engagement. It also permits the intensification of work. Digital technologies enable managers to more closely monitor and manage workflow, with workers under pressure to increase productivity by consistently working at higher speeds and to tighter deadlines, and to rapidly alter schedules to capitalize on opportunities and avoid wasted time.[59] Little wonder that increased demands on our time and attention, with few opportunities to disengage and relax, increase stress levels.[60] Rather than acceleration technologies freeing up time, they fill time and create 'time scarcity' and the experience of being rushed or harried, with there not being enough hours in the day.[61] In the mid-2000s, 35% of Americans reported that they always felt rushed, with almost half saying they never had time on their hands, with leisure time reduced and more hectic.[62]

This is not helped by our fear of missing out on something, which is noticeably intensified by a heightened awareness of just how much is happening thanks to all the platforms we are connected to. There is so much we could do, and we have the means to do it, how could we be content with sitting idly by? The result is we spread ourselves too thinly, doing everything hurriedly and little well. Often our various devices feel like they have become, in Carl Honoré's words, 'weapons of mass distraction'.[63] He suggests that, rather than being in one moment, we're now often in two or three at once, with all the attendant cognitive dissonance that can entail. Time that used to be relatively free of work, such as the commute, or evenings, weekends, or holidays, has been colonized via work and social pressures to engage and interact. We can move, connect, and accomplish more today than ever before. But because we live with others – in the same home or not – and work with others (even if we're self-employed), our responsibility and obligation is to contribute to the broader process of acceleration.

A key consequence of this increase in the pressure to respond, matched with the growing capability of digital technologies to reduce response times, is a new ongoing calculation on the part of those who design devices and services: if more and more humans and machines are engaging in an ongoing back-and-forth to move, connect, and accomplish tasks, it makes increasing sense to automate as many systems as possible. There are limits to what humans can achieve, after all. Better to use digital accelerators to expand the response loop. Better still to remove humans entirely, wherever possible. The result, which you will recognize, is a gradual decline

in human-operated information processing and a related increase in
the number and range of online-only and automated information
processing. Now you *must* renew your driving licence online. Now
the only way to purchase flights is on a website. Now the range of
actual shops you can pop into – and the selection of products they
have in stock – declines (and when you do purchase something you
pay at a self-service checkout). On phone calls to companies and
institutions you can now interact with automated menu systems.
In short, dependence on digital accelerators expands, even to the
point where you will not be able to participate in everyday life
without going online *numerous* times each day. What began as a boost
to efficiency, and of course a boon to profitability, has become a
serious barrier in the way of establishing a balanced digital life.

Some observers have questioned this move towards acting and
responding in such a timely manner. The emphasis on speed and
instant reaction means there is no time for reflection, contemplation,
slow rational deliberation, considered answers, or affect and emotion
in decision making and response.[64] As Robert Hassan notes: 'Users
are compelled by the momentum of the now. Control in this
context is almost impossible: take your time and you lose the sale,
suffer a drop in efficiency, or miss the "valuable" connection.'[65]
Compressed time for thought and action means that actors, such
as company workers, government officials, and city managers, have
to fall back on either learned routines or established unconscious
cognitive biases,[66] or come to rely on automated systems.[67] Family
and friends become hustled into decisions and actions that they
might not have taken with time to reflect. Acting in real-time
erodes our choices and reflexive and meaningful action, and limits
alternative and creative interventions.[68] In other words, *kairos* (the
right time to act judiciously) is trumped by network time and the
need for immediate response.

Many are also concerned that the move towards automated
systems to process and respond to real-time data creates technocratic,
top-down forms of governance. Such an approach prioritizes
optimization, efficiency, and black-and-white decision making
as the key bases on which to manage situations.[69] The worry is
that managing systems and infrastructure in a perpetual present
through algorithms creates a disengaged, decontextualized, rote,
rule-based approach that lacks reflection, deliberation, communal

debate, learning trajectories, and framing to local conditions beyond instrumented parameters and metrics. Living and working like this fails to take account of the wider effects of culture, politics, policy, governance, and capital that shape how everyday life unfolds. It overemphasizes the present at the expense of learning from the past and planning for the future[70] and erases the frame of duration and trends.[71]

To take the management of cities as an example, some commentators argue that smart city technologies seeking to create order, control, and timeliness actually undermine the very nature of 'cityness': their messy, emergent, qualitative experiences; their anonymity and serendipitous encounters.[72] Such technocratic forms of governance generally run counter to democratic politics, with real-time computationally mediated management excluding meaningful public participation in governance, bypassing the creative, political, and messy role of people in shaping their own environments.[73] There is more to worry about regarding acceleration than our personal wellbeing.

Of course, it is important to recognize that these issues do not affect all people in the same ways. Some people are far more snared up in acceleration and issues of time poverty, time pressures, and temporal fragmentation. While the average working week has been relatively stable for a few decades with a slight upward drift, there has been a marked polarization in patterns of work hours, with some people working much longer hours and others who work part-time.[74] Additionally, some people have autonomy over their working hours, whereas others have little leeway or flexibility. Some will be working 9-to-5 Monday to Friday, others will be working shifts and at unsocial hours. Lower social status groups tend to have less autonomy and are more likely to be working irregular hours. Some types of work are much more digitally mediated than others, with those workers using computing devices disproportionately feeling their effects. And some people cannot obtain work, or enough work, being un- or under-employed.

Men and women both experience the effects of acceleration, but there is much evidence that they do so to different extents. As Judy Wajcman details, women are more likely to be managing workplace demands along with more domestic and caring work.[75] As a result, working women do more combined paid and unpaid labour than men or mothers who do not undertake paid work. This particularly impacts single mothers and mothers in dual-income households. Single mothers have little discretionary time and when they do have leisure time it is more likely to be spent with their children or interrupted by them. A key issue for dual-income households is managing and coordinating schedules, especially when partners are working at different times and these also differ to their children's timetables, leading to a sense of being constantly harried. Time scarcity here can be more to do with a lack of synchronization than a lack of time. While digital communication can help, it also causes more interruptions and ever-shifting diaries. The demands of the always-on/everywhere-available work culture also create more friction for women and place them at a disadvantage in terms of career advancement. While both men and women experience work-family spillover (dealing with work while at home), women are more likely to experience family-work spillover (dealing with family issues while at work). Meanwhile, while adults might feel harried by their use of connected devices at times, children are often using them to fill time with entertainment (television, movies, games) and communication (messaging, social media).

Moreover, while everyone is ensnared in a digital world, not everyone has equality or equity in levels of access to networked devices, and there are also variances in who chooses to be connected. For example, not everyone can afford an iPhone or a data plan or smart appliances; while a couple of billion people have smartphones, five billion do not. The use of the internet by people over 65 is significantly lower than younger age groups. Not every employer or job requires employees to always be on a digital leash. Ubiquitous and pervasive computing is more prevalent in the developed than developing world. Not everyone has the strong digital literacy skills needed to use a diverse range of technologies proficiently. In other words, there are social, geographical, and literacy divides across populations.

Negotiating acceleration

One of the most interesting commentators on acceleration is Hartmut Rosa, a German sociologist and political scientist. Rosa's argument, laid out in his book *Social Acceleration* (2015),[76] is relatively straightforward. His focus is first of all on 'technical acceleration', which refers to the technological breakthroughs we've charted in this chapter. He then argues that technical acceleration feeds into and propels 'acceleration of social change' and 'acceleration in the pace of life'. Technical acceleration needs to be at the centre of things for Rosa: in less than two centuries, humans have connected themselves with others in revolutionary ways. This place, *here*, becomes ever closer to *that* place, there. That place *there* and its problems — its civil war, famine, or swine flu — is almost within touching distance. We can watch a distant war or revolution unfold live in the comfort of our home or using our $100 phone while on a train or bus. Dramatic news events break almost instantaneously. It can almost *feel* as if the whole planet is standing still.

The feeling of all this is an important component to emphasize for Rosa. He connects the experiential dimension of technical considerations, such as developments in transportation and communications technologies, to a broader speeding up of social change and an overall increase in the pace of life. Politics rapidly moves on; new cultural mores appear and are replaced at an increasing pace; innovations are diffused quickly and then overtaken by new ones. Rather than the modern person enjoying a sense of independence and control, Rosa suggests that acceleration gives rise to a 'feeling of a loss of autonomy that is manifested in the disappearance of any possibility of control and the erosion of opportunities to shape one's own affairs'.[77] At the same time, there is 'immense growth in the number of possible decisions'[78] we all need to make, which amplifies the sense of pressure we feel. Digital accelerators, in particular, enable us to accomplish tasks, but that, in turn, expands the quantity of responses needed, which leaves so many of us slashing and swiping away at our screens, hammering away at our keyboards, or clicking our way through numerous websites each day. That feeling of a loss of autonomy might not entirely erode 'any possibility of control', as Rosa suggests; nor does it kill off all 'opportunities to shape one's own affairs'. But the pressure is real. The squeezed autonomy, or perhaps we could say the growing constraints on autonomy, is characteristic of our times.

Without focusing on digital life per se, another way of coming at all this is offered by Martin Seligman and his co-authors. Their view, presented in the book *Homo Prospectus*,[79] is that humans are healthiest – less prone to suffer anxiety or depression – when they can project and evaluate future possibilities. We are *homo prospectus* in nature they argue: always looking forward, keen to make some sort of plan, sketching out options. We have, they say, 'powerful abilities for prospective imaginal simulation; we can mentally project ourselves into temporally distant hypothetical situations'.[80] We use our neocortex to pursue ways of negotiating others and to imagine how we can weave our way through the world by anticipating how 'pragmatic contingencies' will be played out.[81] Acceleration enters into the frame here, especially because digital life enables us to accomplish tasks in new and rapid ways, which therefore means we are invested in getting appropriate responses and anxious to see them arrive. If we are *homo prospectus*, then perhaps one reason we have taken to digital life is precisely because it connects with the forward-looking components of our brains.

But following Rosa's argument about the loss of autonomy, it also seems to us that acceleration ruptures our forward-looking possibilities. To take an obvious and immediate example, you have probably used digital technologies, perhaps a service such as WhatsApp, to engage with others around planning an event or a family gathering. You throw messages around, suggest options, send links to websites, share maps, discuss opening hours and entrance fees, and so on. WhatsApp (or whatever service you use) is great for this. 'Ping, ping, ping': tasks are accomplished. A plan takes shape. However, although you are doing exactly what your *homo prospectus* brain is happy with, digital services like these also engender anxiety while you wait for responses. Because you know your sister can reply at any moment, you are uncomfortable with making decisions until there is certainty. You are hesitant. You might have authority to decide and just go ahead and act. But digital technologies encourage information-sharing, participation, inclusion. You wait. You hesitate. New information can suddenly appear: an alternative plan, a different venue, a new time.

The same seems to apply to the way businesses behave today. Probably the best example of this is Apple's reluctance to spend the enormous stockpile of cash it has accumulated over the last ten years in particular. Its executives can imagine how to invest it: automated cars, perhaps, or artificial intelligence. But they have been, and they remain, hesitant. A factor here is that digital technologies reduce barriers to entry: as the extraordinary success of WhatsApp demonstrated, a small team of skilled software developers and engineers can combine their efforts in a short space of time and produce a new popular feature or service (supported by proprietary intellectual property in the form of lines of code) that sucks in users and alters extant markets. Various other apps and platforms have emerged in the last few years in a similar way: Airbnb, Uber, and Instagram stand out as some of the most successful.

We look at these outcomes of acceleration and draw the conclusion that digital life creates a 'hesitant present' where each of us as individuals, and also the companies producing the devices and services so many of us are using, are hesitant about the future. It makes *homo prospectus* anxious and nervy. It affects how we act in the present through a short-term, immediate window of responsiveness, rather than a longer-term plan. It complicates economic calculation and conceivably curtails growth in capitalist economies. Acceleration produces a paradoxical situation whereby all of us are trying to figure out how to act, while also being compelled to act – to be responsive. The feeling of a loss of autonomy is combined with a hesitant present. It is a significant feature of acceleration and introduces dynamics that digital subjects – you, us, companies, governments – must negotiate.

This is not the only paradox created through the relationship between acceleration and digital technologies. As Judy Wajcman notes, for example, digital technologies are supposedly meant to save and free up time, yet they appear to create time scarcity and pressures.[82] To deal with these time pressures we turn to digital technologies, yet then blame them for driving such pressures. Digital technologies open up new freedoms and individual autonomy, but they also create digital leashes and less meaningful engagement, and bind us into systems that demand response. New modes of transportation and transmission speed up time for travel and communication, but they also lead to more sedentary and stationary lifestyles (sitting in a car or in front of a screen, and crawling in congestion or waiting in endless delays). We claim to be rushed and harried by connected devices, but a large portion of our day is spent being entertained through them (games, television, social media) and we turn to them when we're bored, lonely, or just want to fill some time. Despite dozens of new domestic technologies designed to make home life more efficient, the time spent on domestic tasks has not lessened and in many cases has increased (in the main because we do them more frequently or the time saved in one task is invested in another). Digital technologies make it easier to connect and organize meeting others, but the work arrangements they enable (always-on/everywhere-available, work-time drift, 24/7 scheduling) make coordinating meetings more difficult. The more devices and apps we use to manage our lives, the more we feel rushed and harried. For Wajcman, these paradoxes reveal that the relationship between digital technologies and time is not one of straightforward acceleration, but rather is more complex, sometimes acting in contradictory ways.

A key ingredient in these paradoxes is the unfolding nature of society in which our interactions and negotiations take place. In particular, the idea of 'responsibilization', which scholars of neoliberalism (the political economy at the heart of how sovereign states work in much of the world) have used to account for the way we live today, is important. Whether the focus is on health or education, individuals in today's societies are now compelled to look after their own selves first and foremost. Governments are no longer supposed to offer too many supports: public investment is constrained by private enterprises that look to offer goods and services via the market. If you can afford healthcare or education, then you should make that investment for your own sake. Better to do so than hang around on a waiting list or rely on receiving one of a few scholarships. Your life is your responsibility. If you have become wealthy, well done. 'If you are poor, you only have yourself to blame,' is the refrain. Responsibilization works on us, encourages us, and even compels us to act in ways that will bring sufficient material rewards.

Acceleration comes into view here in an extremely awkward way. We enjoy accomplishing tasks. We wait around for responses. We can get things done quickly. We move forward, book flights, order goods, fill in forms, sign petitions, receive notifications, and participate in digital life on an ongoing and expanding basis. But we also experience that loss of autonomy; we understand that we exist in a hesitant present, and yet know, too, that we're ultimately responsible for our own fates. Buying the latest device or downloading an update to our favourite app brings us hope that we'll accomplish more, feel better, move forward faster, and in the end achieve what responsibilization requires of us. But the cost of all this can be severe: anxiety, stress, hesitancy, doubt, constant information-seeking, and even constant failure to achieve plans amid a rapidly evolving scene that can quickly leave us behind. And precisely because there is a harshness to responsibilization – precisely because society is now defined by an ethos of personal success and personal failure (you win or you lose and you shouldn't look to blame others or expect them to help) – the consequences are stark if you do miss out, fall behind, or fail to make the most out of what acceleration offers you.

A constellation of effects

We hope the picture of acceleration we have drawn here is familiar to you. Like the anecdote we used at the beginning of this chapter, we imagine you will recognize the pressure that acceleration engenders. Missing your stop while jabbing away at your screen might not really matter today, so long as you haven't travelled too far and can take another train back to your station. Hopefully you don't miss a crucial meeting. Fingers crossed this isn't a regular occurrence: you haven't been warned about this before, have you? Perhaps for you the specifics of that anecdote don't apply, but the pressure is familiar in some other sense. Maybe there's a feeling that you cannot escape the digital leash; that your use of devices and services creates a persistent sense of being harried, stressed, anxious, doubtful, and/or hesitant.

The important point, we feel, is to recognize that digital accelerators are becoming a central part of our lives, albeit in diverse ways, and that we're now exposed to a constellation of effects. Some effects are positive, for sure, but the downsides can be serious. The task for all of us, therefore, is to work on figuring out ways to strike a balance: enjoying what fast computing offers us but slowing it down when necessary. We cannot see any other way to get that balance right without starting to think through what slow computing can offer all of us.

In later chapters we will elaborate further what we mean by slow computing, from both an individual and a collective perspective, and encourage you to think some more about how it might become a central part of your digital life. But before we can do so, we need to work through another set of relevant issues regarding digital life today. If acceleration is one key element to consider, extraction is the other and so we turn now to consider what this is all about and why it matters so much.

3

Monitoring Life

Wednesday. It's 10.30am and you're heading out to grab a coffee at your favourite spot, Lou's Café. Swiping your staff card, you exit the building, passing the security booth, glancing up at the cameras watching you leave. You tap your phone to hail a rideshare cab, which arrives just a minute later and takes you toward the town centre. At the coffee shop you pay with a tap of your debit card, collecting points with your loyalty card. While you wait for the coffee, you notice a screen near the milk which flashes an ad directed right at you, your name at the top: 'Make warts a thing of the past. Get 20% off our cream at Chem-Care. Scan the QR code for an in-store voucher.' That was embarrassing. You wish these ads were less intrusive. Checking your phone out of habit – why is this coffee taking so long to arrive; you've been waiting a minute already – you see another ad for Chem-Care, which is now offering a two-for-one on wart cream if you tweet about Lou's Café.

In today's digital world, the line between your public persona and private affairs is blurring. The scenario above is an exaggeration: chances are, an imaginary company like Chem-Care wouldn't actually direct such a personal ad in public like that, although tying your purchase in one venue to your location in another is already happening. What you think is your private business – you have a couple of warts, they're annoying, you're bothered by them, and you'd rather not advertise this to others – is now a business opportunity that algorithms can identify and use as the basis for steering your future purchases in a specific direction. Massive investments are being made right now by all sorts of companies to capitalize on and monetize these sorts of opportunities. Whether it is information based on your online search activity or the places you frequent when you're out spending money, you're now just one of billions of other targeted humans.

Digital life demands and creates data shadows and footprints made up of trillions of our individual responses to the prompts and invitations playing out on our devices, the services located within them, and the world at large. These responses form what John Cheney-Lippold calls 'reserves of data'.[1] The data are created when we log in at work and across the various digitally mediated tasks we perform. When we pass a security camera and our face is scanned and recognized, or not. When we hail a cab and complete the transaction. When we arrive at a coffee shop and our phone 'signs in' to the Wi-Fi without us knowing it. When we swipe a credit or loyalty card. When we tap here or there on a device. When we browse the internet during our commute. When we walk past a people-tracking sensor that scans the MAC address on our phone or a traffic camera than scans our licence plate. When we ask Spotify to stream music in the morning. When we choose to watch all of *this* show on Netflix but abandon another without finishing episode one. When we talk to Alexa or Siri or our smart TV. When we shop, sign an online petition, like a Facebook post, or write a tweet. We're making data all day long, when we're online and even when we're not. As Cheney-Lippold puts it, we *are* data.[2]

But does any of this matter? The Chem-Cares of today probably don't really know who you are, unless you have registered personal details with them directly. And they're unlikely to have the time or resources to drill down to such a point where they know your name or individual health history. Yet their futures – their ability to shift product, their stock prices – are tied into knowing more about you and finding ways of getting you to spend money with them. What you do online, where you go, how you move through a city: herein exists marketable knowledge that a Chem-Care or a Lou's Café can make use of. Other companies though – those that help the likes of Chem-Care target potential purchasers – do care about individual identity and try to create detailed individual personal profiles by aggregating data from dozens of sources. For example, Acxiom – a data broker recently acquired for $2.3 billion by Interpublic Group, one of the four largest advertising conglomerates in the US – had data related to 2.2 billion people worldwide in 2018.[3] In 2012 it had about 1,500 data points per person (the company has detailed dossiers on 96% of Americans).[4] Acxiom claims to be able to provide a '360-degree view' on individual consumers by meshing together offline, online, and mobile data.[5]

These databanks enable data brokers to identify and assess the value and risk of existing and potential customers, both now and in the future, and to 'social sort' them (that is, make judgements and decisions about them) or 'nudge' them to act in certain ways (such as purchasing particular products). We might never have heard of Acxiom or other data brokers, but nonetheless they will have had some influence on an aspect of our lives – determining whether we received targeted advertising or a special offer, or a loan, or a job, or a lease.

Data, data services, and data-driven decision making are not just generated and used by private companies. Government departments and public agencies create huge volumes of public administration and operational data to manage the population and activities in their jurisdictions. They are using commercially generated data and working with private companies to extract insight and value from datasets. This includes using the data to make decisions about service entitlements and interventions. For example, public administration and other data are used to determine access to welfare benefits and identify fraudulent claimants.[6] Data are also used for policing and security, often in ways that exceed what we expect: public reaction to the Edward Snowden leaks concerning the extent to which the NSA, and allied intelligence agencies, were spying on both their own citizens and those in other countries demonstrated how little we knew about the extent of data extraction.

All of this raises fundamental questions about privacy and our rights as to how data should be used. Our accelerated, digital lives are monitored and the data generated used to profile and make decisions in relation to us. Can you – should you – do anything about this? Or, if we are data, are we also now just powerless computable, calculable elements in emerging sets of profiles, traits, and connections? Is privacy and data protection possible in a digital world? The focus of this chapter is the accumulation of reserves of data about every facet of our everyday lives and the value mined from these data. We provide an account of the data extraction process and associated downstream services, highlighting features we think you – like us – implicitly approve of, as well as more sinister and, so far as we and information commissioners and regulators see it, worrying developments.

Data extraction is a fundamental part of digital life today. There are few signs that anything other than an expanded process of harvesting data will take shape in the future. One way or another – from the two extremes of blithe acceptance to radical opposition – we all need to deal with this core aspect of digital life. As with the processes of acceleration, we favour trying to strike a balance: we support creating a digital future that offers the joy of computing without us becoming mere data dupes in the system. The first step towards striking that balance requires greater awareness of the extent to which data extraction and data brokerage and services are taking place.

Data footprints, data shadows

Digital accelerators – devices, software, apps – enable us to move, connect, and accomplish tasks and, in turn, they demand responses. Lots of us are typing, swiping, clicking, and tapping almost all day long as we respond to prompts and invitations or initiate communications that require responses from others. The fields within which these exchanges occur will vary: they can be work-related or focused on family life. They often switch across platforms, from smartphones to desktop computers to tablets running different operating systems, and other digital devices such as smart TVs. You'll be initiating and responding inside Facebook or Instagram, or on websites such as Expedia or Airbnb, or other dedicated software interfaces.

Wherever these responses occur, the providers of digital devices and services have scope to record our actions, storing data on where we clicked, how we arrived there, where we went next. Data can be gathered on how long we stayed on a website or inside an app; how quickly we scrolled to the bottom of a page; how long it took for us to return there; whether we read a full article on a newspaper site or clicked a link and headed off somewhere else. There will be data on our interactions, too: did we like a post, retweet something, share a link, or make a comment? And exactly how do our interactions stack up against our prior behaviour? Our phones reveal where we've been in geographic space, how long we stayed there, and if two phones are constantly in close proximity, who we went with. Smartphones are a treasure trove of data. Apps do not just collect information on app use, but can also request data permissions to access a diverse set of information stored on the device, including email, call, and messaging logs; other app activity; data usage; device and battery information; location; and telephony and wireless details. They can also request permission to access device functions, such as the camera, phone, and stored media, as well personal or sensitive data such as addresses and passwords.[7] Some of the data hoovered up by an app will be vital for it to work, but other data are simply of interest to the developer. Crucially, it is not only individuals who are subject to such an intensive gaze, but also objects, transactions, institutions, and places.[8]

Reserves of data do not only arise from our digital actions. Even when we might like to think we are 'offline' we are still exposed to systems and processes that look to collect data about our lives. When we commute on public transport or in the car, data are gathered about where we (dis)embark or when we enter one zone of the city or another. How you navigate the supermarket might be recorded. At a protest march, your face might be scanned and recorded. Walking through a shopping mall, you might be counted or tracked. Calling the insurance company or bank, the pitch of your voice can be monitored, and your recent purchases analyzed instantaneously to determine whether you are telling the truth. The divide between online and offline is blurred. We are exposed to data extraction in more ways than we might imagine. This 'offline' action – managed by companies and governments – demonstrates how digital life today should be understood to involve extraction, not simply exchange.

In short, whatever we do online, whenever we engage with other people or machines in an app or on a website, our responses can be stored in the form of data. The same now often applies to our behaviour offline. Tasks that were previously unmonitored – what television programmes we watch, the settings of our home heating thermostats, the route and mode of journeys we take across a city, or our basic biometrics and activity such as pulse count and steps taken – are now routinely tracked and traced through 'smart technologies'. Some of these data consist of digital 'footprints': the data that we leave behind through our use of technologies. Other data are the result of digital 'shadows': information generated by others without us necessarily knowing it has been produced, for example a sensor tracking our phone's movement along a street.[9]

It is becoming difficult to take part in daily life without leaving some digital trace of participation. Even if you do not use a credit card to purchase goods in a store, your presence is recorded by surveillance cameras; even if you use an anonymous username on social media, your device details are recorded. All of us routinely leave data trails and shadows in our wake, though we often have little control over their form, extent, or how they are used. These data are easily shared and merged with other data and they can be stored indefinitely, locally on devices or in the cloud.[10] Google and Facebook, for example, store *indefinitely* every bit of data we give them and can do so even after you've deleted them from your apps and folders.[11] For example, Google has a record of every search you've ever conducted across all your devices. If you use an Android phone it has a record of every app and extension you've used, how often and where you used them, and who you used them to interact with. If you use Chrome, they know the history of all your bookmarks; Google Drive, all your files; Gmail, all your emails; YouTube, what you've watched; Play Store, what you've bought; Calendar, your diary; and so on across their product range.[12] Facebook store a full record of the time, date, location, and device of every login; every post, comment, like, share, and message; every friend and interaction; all the applications you've ever had connected to your Facebook account; your movement across websites that have a Facebook 'like button'; information about your device and other apps used; and a host of other data.[13]

Academic researchers call this process 'datafication':[14] the increasing capture of everyday life as a continual stream of data. Datafication gives rise to a form of 'surveillance capitalism',[15] and is used to inform what emerging and well-established businesses do. It hinges on the production of detailed, fine-grained longitudinal datasets, which get joined with others and crunched by data analytics. The extent of datafication can be surprising. For example, even by 2011 the Dutch Data Protection Authority reported that the average citizen in the Netherlands is captured in 250–500 databases, with more socially active people included in up to 1,000 databases.[16] We might also be surprised by the numerous ways our characteristics, behaviours, and thoughts are captured by today's technologies and systems.

To extend this a little further, it's instructive to consider the datafication of our location and movements. Up until relatively recently, tracking the movement of individuals was a slow, labour-intensive, partial, and difficult process.[17] The only way to accurately track your position would have been to follow you in person and to quiz those with whom you interacted. As a result, your movement was largely undocumented, with the exception of specific journeys such as flights. Even if you were tracked, the records tended to be partial, bulky, difficult to cross-tabulate, aggregate, and analyze, and expensive to store. A range of smart technologies have transformed geolocation tracking to the point where the monitoring of location is pervasive, continuous, automatic, and relatively cheap; it is straightforward to process and store data; and it is easy to build up travel profiles and histories.

For example, many cities are saturated with remote controllable digital CCTV cameras that can zoom, move, and track individual pedestrians, with some aided by facial recognition software.[18] Large parts of the road network and the movement of vehicles are surveyed by traffic, red-light, congestion, and toll cameras, many of which use automatic number plate recognition (ANPR) software.[19] Smartphones continuously communicate their location to telecommunications providers, either through the cell masts they connect to, or the sending of GPS coordinates, or their connections to Wi-Fi hotspots. In a number of cities, sensor networks have been deployed across street infrastructure such as bins and lamp posts to capture and track phone identifiers such as MAC addresses.[20] The same technology is also used within malls and shops to track shoppers, sometimes linking with CCTV to capture basic demographic information such as age and gender.[21] Public Wi-Fi can capture the IDs of devices which access the network and can track them between Wi-Fi points.[22] Many buildings use smartcard tracking to monitor and control movement through their spaces. Smartcards are also used to access and pay for public transport. New vehicles are routinely fitted with GPS that enables the on-board computers to track location, movement, and speed. These devices can be passive, storing data locally to be downloaded for analysis at a later point, or active, communicating in real-time via cellular or satellite networks to another device or data centre. Many cars are now fitted with unique ID transponders that are used for the automated operation and payment of road tolls and car parking. Our location and movement have become thoroughly datafied.

The actions based on using these reserves of data provide further evidence that a data grab is at work. Gathering is one thing. How those reserves of data are used is quite another. In a general sense, companies that extract data are discovering opportunities to make large profits by pursuing a collection of strategies. In the immediate term, data about our swipes or likes can be aggregated, packaged, and used to sell targeted advertisements. Facebook and Google are world leaders in this regard. They have rich reserves of data because so many digital users are active within their respective 'ecosystems': they have acquired significant knowledge about what their users like, search for, and possibly even believe. Advertisers desperately want to place links to their products within the visual field of these users. Accordingly, they pay handsome sums. In turn, this increases the profits of Facebook and Google, which boosts their market valuation, and enables them to acquire other companies that succeed in drawing in users and data. Facebook's purchase of WhatsApp and then Instagram demonstrates exactly how this process unfolds. In the shadow of these digital giants, there are all sorts of other companies pursuing the same general strategy: tying content production to data extraction with a view to selling ads. In today's global economy, with its $75 trillion GDP, the digital advertising market is worth around $500 billion. If it continues to grow at 5% per annum until 2025, the market will be worth $775 billion – any company that can access even a small portion of that market could have a viable business.[23]

Using data to sell ads, however, is a short-term strategy. The longer-term play involves using reserves of data in more ambitious ways. Developments such as machine learning and artificial intelligence are the frontlines of technology. Acquiring reserves of data is a crucial factor here: a machine learning operation works best when it has a huge dataset to work with. Little wonder that Alphabet, the parent firm of Google, is active in this space: its machines are learning not only from its unique reserve of trillions of search queries but also from data from numerous other fields of action within its enormous range of services. Building intellectual property is about trying to secure rewards years from now: it is about strategic research and development, not the relatively simple act of selling ads. The task in hand for technology companies is creating devices and services that work today but with a view to developing the technologies we'll be using tomorrow. Data extraction presents a unique opportunity to learn about consumers – whether individuals, institutions, or governments – and identify what sort of architecture of devices and services will bring monetary rewards later. Patenting can secure intellectual property; reserves of data can be used to identify what patents should be developed.

Our data footprints and shadows are fragmented and dispersed, divided across dozens of organizations and servers, but there is a growing industry trying to pull them together and merge them into individual and place profiles. Data brokers allow our histories to be examined with respect to issues such as key demographics (age, gender, race, religion, and so on), household composition (who lives with who), credit and income (financial history), employment (who one works for and what one does), health (ailments and illnesses), education (qualifications), housing (ownership/tenancy status), crime (encounters with law enforcement), consumption (records of purchases), travel (where one visits, how one travels), social networks (who one knows), interests (hobbies, sports), political values, and other aspects of everyday life. The data are aggregated from numerous sources, including the databases held by companies (for example, banks, insurers, supermarkets, credit card issuers), social media sites such as Facebook, smartphone apps, and government databases. Data brokers such as Alliance Data Systems, eBureau, ChoicePoint, Corelogic, Equifax, Experian, ID Analytics, Infogroup, Innovis, Intelius, Recorded Future, Seisint, TransUnion, and others are the backbone of today's 'surveillance capitalism'.[24] Each company tends to specialize in different types of data and data products and services, such as performing search and background checks, micro-targeting advertising, identifying high value customers, predictively modelling individual behaviour, and assessing credit worthiness.[25] They sell the data and their data services to a range of clients, including public sector bodies, banks and financial services, insurance companies, media conglomerates, retail chains, healthcare providers, telecommunications industries, and others. Our data are thus used to profile, assess, sort, and nudge us.

Trading data for services

While our introduction sets up the level of data extraction taking place as an issue of concern – and we most definitely think that it is – we cannot consider data footprints and shadows without pointing out that there are some things to like about all this action. On the surface, profiling appears to be a win-win situation for both customers and vendors: users receive personalized treatment and recommendations (such as targeted rather than random ads) and vendors gain revenue and a stable user base. Moreover, in many cases, subscribers get to use a service for 'free'. In a sense, it's a simple exchange. And indeed it can feel like a good deal, especially if you do not have a lot of spare cash to pay for dozens of digital services.

That these services are useful is also worth remembering. We now have powerful tools for connecting, moving, and accomplishing. Digital services enhance our everyday lives. Crucially, many of these digital services – offered for free, because they are funded through monetizing extracted data – are simply better, more sophisticated, and more useful and entertaining precisely because they are free. Free use invites users, which generates responses and provides a target audience and a supply of data, thereby bringing in revenue and raising funds for developers to be hired to improve things, which they can do precisely because there are lots of users and ample evidence of what users want. There are always features we might not like in a service such as Gmail, say (and those features might remain precisely because they allow Google to learn about us), but so many people use the service because they have found that it is superior to competing services, such as Microsoft's Outlook or Yahoo's free email offering. The surge of users on Twitter, Facebook, Instagram, or other platforms suggests lots of us are willing enough to produce data to lubricate all of the action occurring behind the scenes, so long as we can participate in what's taking place front of house.

Even where use is not free, data extraction permits knowledge about users to be gathered and deployed in ways that sustain business models. In the case of Netflix, data extraction informs the company about how it can satisfy users, or at least deliver content that retains subscribers. The core model of Netflix is to attract and then retain subscribers, so their data extraction systems use algorithms to learn about each of us and bring suitable content to our attention: just when you might dare to think about giving up on it, Netflix's algorithms will be trying to ensure you discover a new series or movie you will like. Judged by growth in its subscriber base – an impressive but still minuscule 151 million households in 2019[26] ('minuscule' because at least another a couple of billion households around the world with internet access have yet to sign up) – its data extraction is playing a part in satisfying its customer base (admittedly, never entirely but at least sufficiently).

Data extraction also underpins growth and expansion of other flagship digital businesses. Amazon stands out here. Now the world's second largest retailer after Walmart,[27] it is a leader in using customer and purchase history data to drive targeted recommendations which convert into sales. Its growth into a giant enterprise is directly attributable to its ability to use its data analytics capabilities to extract insight from the data we provide simply by viewing products on the site and buying things.

Beyond swapping data for services provided by digital firms, we also swap data with non-digitally native,[28] traditional companies, such as utility or transport providers. In return, we receive a service which we hope will be more responsive, customizable, and inexpensive. Smart meters in homes, for example, do not just allow an energy company to plan and operate their supply more efficiently, effectively, and sustainably; they also provide information to customers so they can monitor their own usage and plan around lower tariff periods. In the supermarket, loyalty cards invite us to swap data about purchases for targeted discounts. Then there are service providers such as hotel chains and airlines that use customer relationship management (CRM) systems to try and build personalized relationships with customers by using data on bookings and preferences to offer loyalty rewards, special offers, and upgrades. All of these systems enable service providers to prompt staff and systems so they can appear to 'know' you when you call, arrive, check in, or use their websites.[29] The trading of data is thus framed as improving customer experience, while enabling the company to leverage competitive advantage, be more flexible and opportune, and reduce risks, costs, and operational losses.[30]

With respect to public services, we frequently swap data for improved efficiency, optimization, competitiveness, security, safety, and so on. Systems such as government websites and e-government, surveillance cameras and security systems, centralized control rooms, coordinated emergency response, and intelligent transport systems and sensor networks aim to use data to provide personalized services, fulfil entitlements, reduce fraud and crime, address health and safety, manage congestion, tackle social and environmental issues, direct public investment, and help government do more with less finance. It doesn't always work out this way, but all of these systems can improve our quality of life and wellbeing and possibly even reduce taxes. Moreover, if data are aggregated and made open, they can also increase transparency and accountability or encourage social entrepreneurship around community development or engender new market opportunities (such as creating civic apps). Providing data for these benefits seems like a fair exchange.

Our experience of digital life, then, is tied up with data extraction. In turn, data extraction makes it possible for companies and government to present features to us and make invitations and prompts that sometimes appear creepy and worrisome, but also sometimes smart and appropriate. Certainly, the algorithms operating behind the scenes and working their way through reserves of data about us have successfully produced facets of digital life that a great many of us enjoy. For all that we can – and should – identify more worrying consequences of data extraction, which we now move on to consider, the joy of computing today is intrinsically linked to deriving diverse forms of value from data: not just economic but also social value for the sake of improving society.

The drawbacks of data extraction

As we have noted, data extraction creates numerous overlapping trade-offs. It is part of a business model that has functioned well when judged only by the proliferation of digital devices and services underpinned by the potential for providers to gain value from data. But we cannot judge digital life on these terms alone. We need to take a critical stance on the construction of vast reserves of data and the way companies and governments work with, and on, data. For the purposes of the discussion here, we want to suggest that data extraction is problematic in four core ways, which we will discuss in terms of (i) privacy, (ii) difference, profiling, and sorting, (iii) governance and politics, and (iv) production.

Privacy

Though a contested term, privacy is about the selective revealing of ourselves to others and the protections regarding the accessing and disclosing of personal and sensitive information about us.[31] At an individual level, privacy consists of a number of different facets:

- identity privacy (to protect our personal and confidential data);
- bodily privacy (to protect the integrity of our physical bodies);
- territorial privacy (to protect our personal space, objects, and property);
- locational and movement privacy (to protect against the tracking of our spatial behaviour);
- communications privacy (to protect against the surveillance of conversations and correspondence); and
- transactions privacy (to protect against monitoring of queries/ searches, purchases, and other exchanges).[32]

Privacy is considered a basic human right in most countries and is enshrined in national and supra-national laws. Breaching privacy can have a number of effects on the emotional and physical wellbeing of individuals, as well as opening them up to various pernicious activities such as disclosure, exposure, distortion, exclusion, appropriation, identity theft, and blackmail.[33]

Given the extent of the data extraction underway in the digital age, privacy is seen as being systematically under attack. We are simply under much greater levels of intensified surveillance than ever before. Moreover, we live in much more open and transparent societies than we used to. Information that was previously considered private is being more freely shared, such as résumés (via LinkedIn), family photographs and videos (via Flickr, Instagram, and YouTube), personal and family stories (via Facebook and blogs), and personal thoughts (via Twitter, chat rooms, and online reviews).

What might have been shared with a handful of people (family, close friends, employers) in limited forums (the home, a local bar, a HR office) is now being globally broadcast for anyone to view.[34] Of course, not everyone is free and easy with personal data, but even these individuals can be tagged and included by others. You might not have an Instagram account, but your photo and name might nonetheless be shared there hundreds of times. And even when you do deliberately share information, you do not necessarily anticipate that the data might be packaged, sold on, and worked on for reasons that have nothing to do with its original purpose. And yet this is happening regularly.

Large numbers of apps do not have privacy policies that users can view and accept. For example, the European Data Protection Supervisor reported in 2014 that 39% of the most popular apps had no privacy policy.[35] Likewise, a comparison of 110 apps in 2015 found that 33% of iOS apps and 25% of Android apps had no privacy policies.[36] Given that in 2015 there were over 1.5 million apps in the Apple App Store and 1.6 million in the Google Play Store, a very large number of apps lack privacy policies.[37] As we have noted, a common worry is that many apps practise 'over-privileging', that is seeking permission to access more data and device functions than they need for their operation alone.[38]

Moreover, notice and consent – considered the cornerstone of privacy and data protection – have been significantly weakened over the past couple of decades. Where apps do have privacy policies as part of their terms and conditions, they can be difficult to understand given their use of both legal language and vague terms, their length, complexity, and declarations about changing future terms unilaterally.[39] And configuring the privacy tools within device settings is not always intuitive to non-technical users. Given the volume and diversity of all the digital systems we interact with, it is onerous to try and police our privacy across dozens of platforms and services, to weigh up the costs and benefits of agreeing to terms and conditions without knowing how the data might be used now and in the future, and to assess the cumulative effects of our data being merged with other datasets.[40] As a result, Daniel Solove, a leading privacy scholar, notes that '(1) people do not read privacy policies; (2) if people read them, they do not understand them; (3) if people read and understand them, they often lack enough background knowledge to make an informed choice; and (4) if people read them, understand them, and can make an informed choice, their choice might be skewed by various decision-making difficulties'.[41] Consent thus often consists of individuals unwittingly signing away rights without realizing the extent or consequences of their actions.[42] The consequence is that 'privacy policies often serve more as liability disclaimers for businesses than as assurances of privacy for consumers'.[43]

Beyond the web and apps, notice and consent can become an empty exercise or indeed be entirely absent. Associated activities such as data mining and repurposing are often covered by catch-all disclaimers, along with the right to unilaterally change terms and conditions without notice, effectively disenfranchising individuals of choice, control, and accountability. In the case of some smart city technologies, there are few ways to even ask for notice and consent and therefore no choice but to accept surveillance. For example, facial recognition cameras, automatic number plate recognition cameras, smartphone MAC address or Wi-Fi tracking, and connection to the internet of things all take place with no attempt at consent and often with little notification (though there may be notice in the form of information signs in the area being surveilled, or on associated websites but not on the physical site itself). Moreover, there is no ability to opt out[44] other than to avoid the area, which is unreasonable and unrealistic. There is no sense in which you can selectively reveal yourself; instead, you must always reveal yourself.[45] Moreover, if you are unaware that data about you are being generated, then it is all but impossible to discover and query the purposes to which those data are being put.[46] Furthermore, opting out of the tracking and trading of data is difficult. In 2014, journalist Julia Angwin identified 212 data brokers operating in the US that consolidated and traded data about people; only 92 of these allowed opt-outs (65 of which required handing over additional data to secure the opt-out). She also identified 58 companies that were in the mobile location tracking business, only 11 of which offered opt-outs.[47]

As such, there is little doubt that the concept of privacy is changing, challenging both social and legal expectations. For some, the notion of privacy is largely dead.[48] It is seen as difficult to maintain in practice, it curtails user experience, it is an economic hindrance, the majority of people do not seem 'to mind being mined', and if you have nothing to hide what is the problem with data being open and shared?[49] Data are the means of exchange and in a world where many services are free, then the price for convenience, enjoyment, and a tailored service is privacy. As the saying goes, 'if the service is free, then you are the product'. Users largely know and understand this and consent to it via end user license agreements (EULAs) – which we 'agree' to when we sign up to a service. In this view, whatever happens to data after it has been collected is up to the company: if they can make it work, if they can sell ads or use data on swipes or clicks to develop new intellectual property, then so be it.

Counterarguments point out that contemporary life is necessarily digital: this means users have no option other than to create digital footprints and shadows. In effect, we have no choice other than to sign a EULA or to forfeit notice and consent. What appears to be a relationship of exchange is actually emerging from coercion.[50] This imbalance in the exchange, privacy advocates argue, needs to be redressed to ensure that people regain the rights they previously enjoyed. Privacy is seen as a right that has to be protected: it is foundational to an informed and reflective citizenship, to freedom of expression, and is therefore a central element in liberal democracies.[51] Without privacy, our individual and collective liberty and future are potentially at risk: societies who have implemented mass surveillance have not been socially progressive. This is the logic that drove the implementation of the GDPR in the EU (though it only applies to EU countries).

Difference, profiling, and sorting

Following on from questions of privacy are questions as to how data are used, not just from an economic perspective of monetization, but also how they interact with and sometimes complicate (and even amplify) cultural differences and how people are treated in differential ways. There are three key issues of concern here.

One is about the way humans design data extraction processes, often imperfectly. As we have noted, data are accumulated in enormous reserves controlled by all sorts of actors, including firms and governments. Objectives vary. Microsoft might want to use data to develop new software. Facebook want to sell targeted ads. Government might want to identify potential criminals or terrorists. In the process of making the systems to enable data extraction and mining, designs have gone awry in problematic ways. In one emblematic case, Google Photos introduced a facial recognition feature that identified black women as gorillas.[52] In another, Google Search has been critiqued for reinforcing racist stereotypes and racism through its search results.[53] Elsewhere, predictive policing technologies in the US encourage and reinforce racial profiling, which disproportionately directs police attention toward black men.[54] Lacking critical scrutiny, implicit biases are put on display with nefarious effect. As Safiya Umoja Noble discusses in her book *Algorithms of Oppression*,[55] a central and problematic feature of digital life is the lack of diversity among technology designers and coders. Moreover, these designers and coders are prone to claim the discriminatory results of their designs 'lie entirely exterior to the encoding process'.[56] Rather than tackling system biases, too often the tech industry tries to hide them under layers of digital denial, thereby allowing biases to propagate further. In short, too many facets of digital life fail to take seriously the significance of human diversity and the consequences of inequitable treatment within technological systems.[57]

Beyond design, a second issue concerns the contingent way data extraction meets up with the real world. Companies or governments promote the use of algorithms and automation, for example, as a way to reduce operational costs or improve the accuracy of information processing. But even when biases are considered, and efforts made to reduce them, digital devices and services confront an unequal and diverse world. They are inevitably experienced by different types of users in unexpected ways. There is no way to avoid this. No form of artificial intelligence, even one that has been designed carefully, can anticipate how users in all their diversity will encounter what they see in front of them. So when software is used to predict criminal activity, say, or when data extraction via a site such as Facebook is used to shorten a large list of job applicants, unavoidable biases and mistakes will play out. The issue here is that cultural difference cannot be reduced to simple correlations between data. Your Tinder or Match.com profile can only 'see' one part of you; there is more going on – your personality, tastes, or faults as they are played out in one context to the next – that digital devices and services cannot know. As much as we *are* data, we are also much more than data can ever truly, accurately capture. Moving through space and time, humans demonstrate flexibility, an unmatched ability to adjust, and extensive forms of intensive intelligence that even the most carefully designed computational systems will fail to anticipate.

Third and most problematically, the data deluge produced through extraction is open to 'dataveillance', social sorting, and predictive profiling. Dataveillance is a form of surveillance enacted by sorting and sifting datasets in order to identify, monitor, track, predict, and regulate people and their actions.[58] Social sorting consists of categorizing people based on their data in order to treat them differentially, with some receiving a preferential status and others marginalized and excluded.[59] For example, a company might socially sort customers based upon where they live, with those living in poor neighbourhoods receiving different offers, prices, or credit. Predictive profiling uses stored data to make inferences about people and to then treat them in a particular way. For example, a company might construct a profile of a person and then try to predict their credit risk: how likely they are to be able to meet payments, their projected lifetime value if they remain a loyal customer, and how likely they are to move their custom.[60] Using this predicted profile the company can prioritize its attention and resources, usually focusing on high value customers, perhaps redlining those deemed unprofitable, not profitable enough, or high risk. Or companies might offer varying pricing across customers based on profiles: common in the insurance industry, this is now being rolled out to other sectors. For example, the retail sector has begun to implement systems where shoppers in the same store, or visitors to the same website, pay different amounts for the same product based on their customer profile.[61] The aim is to leverage optimal spending in the store's favour by usurping 'the entire value surplus available in the transaction by pricing goods or services as close as possible to the individual's reservation price'.[62] These practices are not about implicit biases, rather they are explicit attempts to identify particular kinds of people and treat them differentially.

Governments and public agencies can also undertake profiling to determine the services and entitlements of citizens and also determine if any state interventions are required. For example, initiatives might seek to detect children who might be at risk or in need of additional supports, or uncover people who are living in the country illegally or are not paying all the taxes they owe, or identify criminal offenders in the case of predictive policing.[63] Such initiatives might also try to determine which areas are in need of targeted responses to particular issues, such as education and labour programmes in places with high social deprivation, or investment in infrastructure or tax breaks for development.

In the case of predictive profiling with respect to individuals, such practices can constitute what has been termed predictive privacy harms.[64] Some inferences produced through predictive profiling can constitute personal information; for example, predicting sexual orientation by looking at friend patterns or posts on social media, or religion based on what places a phone has been to (such as a mosque), or pregnancy based on shop purchases. Such information might be sensitive and if acted upon could have consequences. For example, if the sexual orientation of somebody who is still in the closet is used to send advertising to the family home or into a timeline on social media on a shared computer, then it could cause personal harm. Or if the person is living in a country where being gay is still illegal it could lead to physical violence or criminalization.

The worry is that predictive profiling is leading to what the former chairperson of the US Federal Trade Commission, Edith Ramirez, calls 'data determinism'.[65] We are not simply profiled and judged on the basis of what we have done, but on a prediction of what we *might* do in the future. Data determinism is most clearly expressed in predictive policing, where analytics are used to try and prejudge where crimes might occur and who might commit them. A number of US police forces are now using predictive policing. For example, the Chicago police force undertakes both area profiling to identify patrol routes based on patterns of crime and individual profiling that tries to predict who is already or may become a criminal by examining the social networks of those arrested via phone records and social media.[66] Those identified are labelled 'pre-criminals' and are visited by police, not because of what they have done, but because of what an algorithm predicts they might do. In such cases, our data shadow does more than follow us; it precedes us. The potential for civil rights infractions here is very worrying, with people being treated as if there is *evidence* that they have done something wrong rather than simply inference and conjecture. This is compounded by the fact that profiling systems are biased in various ways, working to further discriminate against particular groups. Rather than creating a fairer society, data-driven profiling often deepens inequalities and injustices.

Governance and politics

Social sorting and predictive profiling are two ways in which the governance of society is changing in the digital age. Many commentators argue that the use of digital technologies for the purposes of governance is shifting the nature of 'governmentality' – that is, the underlying logics and mechanisms as to how governance is organized and works. The contention is that governance is becoming more technocratic, algorithmic, automated, and predictive in nature.[67] Technocratic forms of governance presume that social systems can be measured, monitored, and treated as technical problems with technical solutions, rather than solved through other mechanisms, such as regulation, policy, programmatic interventions, social partnerships, or community development. These technocratic systems are underpinned by large-scale data extraction and often work in automated, autonomous, and automatic ways with limited human oversight or appeal.

The effect of technocratic systems is to shift governmentality from disciplinary forms of regulation towards social control. A key aspect of disciplinary governmentality (related to policing, regulation, and workplace conduct) is that people know they are subject to monitoring and enrolment in social and workplace regimes that monitor and reward them, and thus they self-regulate their behaviour accordingly to avoid incurring penalties. Technologies such as surveillance cameras are disciplinary in nature, designed to make us act appropriately for fear of being witnessed transgressing and subsequently punished (even though we know that the feed might be unmonitored). In control systems, we are always being monitored and processed, with our behaviour directed explicitly, or implicitly steered or nudged, rather than being disciplined or self-disciplined. For example, the performance of a worker at a supermarket checkout used to be monitored in person by a supervisor and/or later by a camera; now it is measured by the scanner and till itself: if a worker's scanning rate drops below a certain level, they will be told to speed up. Their behaviour is constantly being modulated.

Data extraction alters governance in other ways. Once users of digital technology know their responses will be recorded and stored, they become aware that their actions are subject to surveillance. Data extraction tweaks the parameters of extant forms of self-governance. Social participation already occurs in ways that reflect how individuals govern their actions with respect to social norms (we police our behaviour in relation to social expectations and fear of exposure to any disciplinary gaze and subsequent punishment). Data extraction amplifies the disciplinary component of self-governance because actions performed online *are* tracked and conceivably presented as evidence against us. You might read a book about the Iraq War and discover the name of a radical cleric whose biography seems interesting: but do you then go online and search for his name? Or do you pause and imagine that such searches are tracked not only by Google but also by the NSA, which actively monitors the online world and looks for users that it deems to be at least 51% 'foreigner',[68] which gives the agency legal authority, according to US law, to track you.

More generally, now that political life occurs in the context of data extraction, enjoying the freedom to think and using that freedom to ask questions about how the world unfolds entails an ongoing negotiation of surveillance across an institutional matrix of data-grabbing activities. Workplaces, libraries, internet service providers (ISPs), governments, and the creators of the online platforms we need to accomplish tasks are all potentially watching what we do. Even when these agents of dataveillance are supposed to be regulated by national government (or in the case of Europe, an EU-wide directive), we cannot know what they are, and are not, recording, or how they are processing and acting on those data.

The upshot is a growing sense across society that the use of digital devices and services is exposing us to experimental forms of governance that we cannot control, nor even know much about. There is every reason to believe that revelations regarding the extent of surveillance pursued by the NSA are the tip of the iceberg. Still out of sight are numerous other forms of dataveillance that interact in as yet unknown ways with our self-governing selves. Data extraction architectures are becoming woven into our mentalities. In this emerging 'data-mentality' relationship, we look out on the world and know that accomplishing tasks will require moving through online worlds and adding to reserves of data. We immediately sense the existence of data extraction and make decisions about how we should self-govern our actions in relation to the diverse ways we are governed within the institutional matrix.

What search engine should we use to find information? On what site will I book a flight or order a book? Which newspaper site will I visit? On what device should I sign a petition? Using which email address will I sign up to hear about an upcoming protest? Answering these proliferating questions is often impulsive, almost without thought. But as we develop our data-mentalities in relation to each new revelation about experiments, or our growing awareness of data extraction, we undoubtedly give these questions more thought. And when we answer them, one way or another, we locate our actions in chains of responses, thereby enacting (making concrete, albeit in ones and zeroes) that precise performance of our data-mentality. And of course, because we spend all day being digital, constantly interacting with multiple digital devices and services, we repeat this process all day long. The problem it poses for our engagement with governance connects with the possibility that data extraction is leading to consolidation among digital providers: the possibility emerges that we have no latitude for independence when it comes to our data-mentality decisions, with all of our responses simply adding to one of a shrinking number of enormous reserves.

Data extraction by companies and governments also alters political life. In ostensibly democratic societies, data extraction raises the possibility of entirely new calculations about how political life can be changed and conceivably manipulated. The controversy over Cambridge Analytica's (CA) use of 50 million Facebook profiles is emblematic of these new calculations.[69] In this case, the data of Facebook users was analyzed to identify openings where certain types of political ads could be targeted that would, from CA's point of view, sufficiently alter voting patterns in the 2016 US presidential election. While Facebook's users know that the company is producing huge reserves of data, they did not necessarily anticipate that their data would be worked upon and crunched according to a specific set of political objectives. Receiving ideologically loaded ads designed to swing an election is quite different to receiving an ad for a pair of shoes, and the stakes are much higher. The code underpinning US democracy was rewritten by this relatively small team of specialists and then installed into chains of responses within sites such as Facebook and other arenas wherever users were active (and producing new data).

Beyond the US, similar activities – including activities led by CA – are now pursued, with unpredictable consequences for how governance processes will occur. An extremely alarming scenario is that data extraction will facilitate certain actors to incite violence: data-infused terrorism, or violent protest and revolution, or even genocide (whipping up a storm and public opinion on social media). Not quite so extreme, but still a concern, is simply the prospect of democratic processes becoming so bound up with data extraction and social sorting that societies become more polarized, perhaps to the point of low-level violence breaking out. There is already evidence that violence, such as school shootings and attacks on democratic representatives, was shaped by what perpetrators had been doing online beforehand.[70] We cannot know the precise layout, but we can be confident of the likelihood that some degree of political manipulation was occurring and conceivably that it affected perpetrators' violent impulses.

Another fear is that deepening datafication and control creep[71] (technologies designed to do one thing being repurposed for another) will inevitably lead to a Big Brother society in which data regimes perniciously shape everyday lives. China's new social credit scoring system, underpinned by mass surveillance and accompanied by pervasive censorship and control of online content and social interaction,[72] provides a sense of what a Big Brother society might look like. The social credit scoring system pulls together 400 datasets from 42 central agencies, 32 local governments, and 50 market actors related to four different domains: public administration (central and local government services), judicial affairs (law enforcement and security, including facial recognition CCTV), social activities (including social media and travel), and commercial activities (including financial affairs).[73] These data are used to assess 'trust', 'reputation', and 'creditworthiness'. The system is presently heavily weighted towards commercial sectors, but has significant social and individual components, and is targeted to be fully deployed by 2020.[74] The calculated scores, the Chinese government argues, are used to enact 'social management' and create a 'fairer society' by ensuring those organizations and people who are deemed most deserving receive appropriate services and tax deductions, and those with low scores are penalized and denied access to some services.[75] Scores can influence access to: government supports and services; market permits; public procurement; economic and financial sectors; housing provision; schooling and university; and what modes of public transport one can travel on.[76] The pro-democracy demonstrators in Hong Kong are well aware of the potential implications of China's mass surveillance infrastructure to their personal freedoms, which is why they have been using encrypted communications, hiding their faces with masks and actions with umbrellas, paying for public transit with cash rather than traceable currency, and even cutting down facial recognition cameras to try and protect their identities while protesting.[77]

Production

The business model underpinning the digital economy pivots on grabbing data about all of us. This introduces a new and problematic dynamic of economic production. When companies move to develop products that rely on the accumulation of data for use in short- and long-term strategies, we become enrolled in their production process. Actual sales remain crucial for manufacturers of all sorts of goods and their employees are still classic 'workers', at least as we have tended to imagine that category of humanity for the last 200 years. But to the extent that digital life involves data extraction, any user of a digital device or service is also now a worker, although we are working for numerous, sometimes obscure companies making profits we know next to nothing about and receive no recompense for our labour. We work in the dark. The apps on our phones sell ads or pass on aggregated data about our lives without us having much ability to know how revenues or profits are accumulated, invested, or distributed. The websites we visit need us to stop by and respond. In doing so, we ship tiny packets of potentially monetizable data into numerous unseen and unknown data reserves: not only, say, the reserves controlled by the newspaper or travel site but also the owners of various services 'plugged into' those sites, including corporate giants such as Facebook or Alphabet but also numerous smaller companies that track or prompt digital users when they surf online.

The relationship between worker and company, producer and profiteer, has been altered in unpredictable ways. Before data extraction emerged to become such a central feature of economic production, the life of a worker was, relatively speaking, transparently related to their employer. Companies or governments moved through cycles of boom and bust and as they did so, workers negotiated in one way or another (collectively or individually) to create an understanding of how their material circumstances would change. Now, in addition to any weekly or monthly payslip we might receive in return for the labour we have expended, we need to recognize that we have also been working online, with few (if any) concrete ways of knowing who has benefitted, but knowing that we have not received a pay cheque.[78] In other words, our data and labour are monetized for no material return. This enables immense profits to be made by relatively small companies – 'small' at least in terms of the number of employees compared to pre-digital companies making similar profits (for example, Facebook, which made $10 billion in 2016, employed around 25,000 people; Ford Motor Company also made $10 billion but it employed around 200,000) – precisely because their 'workers' include non-remunerated users like you or us.

The dispersal of economic production to billions of users – each with multiple accounts and profiles across numerous online platforms – adds a new layer of uncertainty. Once companies pursue data extraction, they cannot calculate how much time users will spend in regions where their responses can generate data. They can tweak code, adjust algorithms, introduce new features, promote certain types of responses, and prompt users to view content. But none of this action can be controlled by any one company. This is a form of dispersed economic production that dismantles control yet requires that data-driven companies create relatively autonomous algorithms which remain alert to sudden surges of activity with a view to temporarily tethering online action to capture data and potentially translate that into revenue. There will be human-driven action here; this is not only a matter of algorithms or bots. When a news story 'goes viral', for example, and there is a surge of users visiting a news or travel site, or spending time within a certain app, human editors or designers will step in to write new content, create new links, or adjust algorithms to flash certain types of ads. The dispersal of economic production requires automation via algorithms but alongside new types of critical human actions that need to be agile, astute, and intimately aware of what might encourage users to hang around and add to the various connected reserves of data. The worrying element here is that, although users like us have some degree of autonomy to decide where we will participate in chains of responses, the basis of data extraction encourages companies to consolidate and grow in size and maximize their ability to ensure we respond within their online regions. This dynamic is cause for concern: dispersed economic production underpinned by data extraction encourages the formation of economic actors that will look to create monopolies and thereby hinder competition.

Another constellation of effects

Returning to your visit to Lou's Café for some coffee, it should be obvious that the only thing Chem-Care (or any other advertiser looking to target you) can predict with any certainty is an unpredictable encounter. You might not mind seeing the ad. You might hate it. It might not embarrass you too much. It could be the final straw. But within this region of the digital world, there lies the possibility that you will tap or swipe or click in ways that make that risk worth taking. Meanwhile, you're aware you have been tracked; the sense within you grows that you're never alone when you're out and about; you're able to grasp that you're producing even when you thought you were simply consuming; and you know, too, that your identity as a Lou's patron can deposit part of your personhood in a category that might attract certain types of political ads.

Coming at us when we're online today is a vast additional array of forces with numerous and diverse possibilities giving rise to another constellation of effects. Of course, it would be churlish not to reiterate that, alongside acceleration, data extraction does bring rewards. But from one moment to the next, and in one context and then another, we can experience those effects in problematic ways. We have outlined four, but like us you can probably think of others.

The rapid pace of change that defines the development of digital life prompts us to carry on and adopt more and more of the latest and greatest devices and services. Like you, perhaps, we will be tempted to rush in, adopt, and adjust to many of these changes. But in the remainder of this book we want to outline ways to avoid that temptation. Slow computing, we will argue, means trying to establish some control amid acceleration and data extraction.

4

Personal Strategies of
Slow Computing

Thursday. It's 1pm. You've been trying to change how you work. You arranged to meet a colleague for lunch by emailing her yesterday, rather than contacting her via WhatsApp today. You've spent the morning working in a focused way, ignoring email and messages, and staying away from social media. Before heading out of the office you've just caught up on those communications in a quick intensive session and placed your phone on silent. Lunch feels a little strange because you leave your phone in your pocket and not on the table next to your plate. You describe the morning to your friend: how you read a book on the train for the first time in years (and how odd that felt when everyone around you was on their phone); how you've purposefully been using a new web browser that hides your IP address; and how you're more carefully curating what you share online. She doesn't seem too impressed, but it has actually felt like a good move to you; certainly, it's been different and, yes, it's been a little bit inconvenient. But it has also felt liberating.

Hopefully this scenario sounds appealing and achievable. We think it starts to set out what a slow computing day might look like. As we pointed out in Chapter 1, slow computing is a means to address the various issues that acceleration and data extraction create. It seeks to do this in a way that prioritizes and protects your needs and interests, and creates public good for society as a whole. Practising slow computing requires individual *and* collective actions: people making decisions to change how they lead their digital lives by considering and changing their own situations, and pooling their actions and drawing on the diverse actions of others.

This chapter focuses on tactics you *individually* can use to fulfil the strategy of slow computing. It is about taking ownership of the issue and being proactive in tackling acceleration and extraction at a personal level. To that end, we set out some interventions aimed at slowing down your computing. We emphasize reconfiguring the time you spend on your digital participation, while curating and limiting the extent to which you leave data footprints and cast data shadows. Adopting these ideas can be empowering because it can give you back a degree of control. But it can mean accepting a degree of inconvenience: the digital world is often arranged in a way that discourages you from practising slow computing. We're not supposed to swim against the tide. Devices, apps, and entire digital ecosystems have been designed in ways that push us to accelerate and expose our behaviour and thoughts to data extraction. Nonetheless, we can take small steps toward establishing a more balanced digital life, and these steps can create a pathway to additional moves further down the line.

It's important to realize you are not alone in desiring what slow computing seeks to achieve. What we think you will discover when you pursue slow computing is that more and more people are pursuing similar practices and sharing ideas, and indeed technologies, which can help you overcome these obstacles. In turn, as you embrace slow computing, you will encounter openings where more collective strategies can be pursued. Chapter 5 shines a light on some of those possibilities for collective action.

This chapter is divided into four parts. We first tackle the question of how we can take some control of our time amid proliferating digital accelerators, and why we might want to do so. We introduce the idea of 'time sovereignty', and emphasize reflection, auditing, and identifying contingencies where you can resist the pressure and temptation to speed up and instead slow down in ways that award more control over your digital life. In part two, we focus on data extraction. We explore how you might engage with, or evade, excessive data harvesting effectively and astutely. We introduce and elaborate upon the concept of 'data sovereignty', which we propose should be the ethos guiding your digital actions. As we point out, this is about capitalizing on the 'sliver of agency'[1] we retain to express our sovereignty over the data we inevitably produce. We set out four sets of interventions: curating your digital life; using open source alternatives; stepping out; and obfuscation.

Part three considers why practising slow computing is hard and why it needs to be constantly worked at. We are under pressure from family, friends, employers, colleagues, and clients to be connected and to use digital technologies. Many of the platforms we use are habit-forming, as well as being psychologically compelling and addictive. We have almost become programmed to use them compulsively. Even if we want to try and unshackle ourselves from certain technologies and practices, it can be difficult to achieve because we are so reliant on them in our everyday lives. Moreover, when the present condition is cast as the natural order of things it can be difficult to imagine and make a different kind of life and society.

Finally, we detail how you might go about contemplating and formulating your own slow computing strategy and the tactics required to put this into practice. We highlight how a slow computing day requires deliberate contemplation about where, when, and how you participate in chains of responses. Whether on the train, at lunch, or in bed, your digital day (and, more broadly, your digital life) can (and, we think, probably *should*) be different. And it can become more balanced if you adopt a critical stance, accept a degree of inconvenience, and embrace slow computing.

Slow it down: practising time sovereignty

> Being Slow means that you control the rhythms of your own life. You decide how fast you have to go in any given context. If today I want to go fast, I go fast; if tomorrow I want to go slow, I go slow. What we are fighting for is the right to determine our own tempos.[2]

As we documented in Chapter 2, we are living in a fast world of 'always-on/everywhere-available' and this has numerous consequences for how we live today. Taking action with respect to accelerating technologies is to take the 'slow' in slow computing seriously. Our call is for all of us to reflect on the way we participate in an accelerated world and on the consequences on our everyday lives, and to consider how we might adjust our practices to slow things down to a more sedate, manageable, healthy, and enjoyable pace. To make *time* for slowing down, for rest, for play, for taking it easy; for escaping the 'never-ending workday' or hectic social schedule. To resist the lure of busyness and speed and to protect our time and attention from systems and devices intent on grabbing our attention. None of this necessarily means 'reaching for the emergency brake'[3] by making a radical reorganization of how we live digital lives. As we've noted, it's all but impossible to step fully outside digital systems. Rather, it is about adopting a range of pace and tempos suitable for the context and our aims. It is about shifting back, when it makes sense, to social and clock time rather than always operating in network time, and to older, slower practices rather than accelerated ones.

Although the speeding-up of social life gives the impression that 'running to stand still' is now required, it is clear that this is not necessarily the case. Often less is more. Slow can actually be more productive and rewarding. There is now ample evidence that structured rest (plenty of sleep, free evenings and weekends, vacations free of work) and strategic scheduling (blocking tasks, minimizing distraction) produce more and better outputs than rushing, multitasking, and working all hours.[4] Speed and quantity do not equal quality. Rather than creating efficiencies, multitasking often means doing two or more things poorly. Often, technology acts as a false friend: while it saves time it also spawns a whole new set of duties and tasks.[5] For example, email makes communication easier but that in turn multiples the number of messages received and accelerates the turnaround time for response.

People who can create detachment from work – who practise structured rest – are happier, more productive, resilient, and less error-prone.[6] They are also more pleasant to work with, improving collaboration and client relationships. Not only is it better for your work, it is better for your health, reducing stress-related illnesses (such as insomnia, migraines, hypertension, depression, asthma, gastrointestinal conditions, and eating disorders) and the likelihood of burnout.[7] In turn, this improves the productivity of companies by minimizing days lost to ill health and the associated costs. There is also an argument that slow makes our lives more fulfilling and meaningful. If you slow down you have more time to enjoy activities without distractions, consider issues more thoroughly, try solutions out, deal with deeper, more thorny problems, and contemplate life.[8] You can do things for their own sake and not for some ulterior motive.

In play here is an ethics of time that requires individual and collective reflection, conceivably enrolling institutional structures, on the nature of 'time sovereignty' in the digital age. By time sovereignty, we mean the power and autonomy to dictate how as individuals we spend our time. While we all have some level of control over our diaries, we are also bound into social and institutional obligations. When managing our own schedules, we have to fit these around or negotiate with other family members, friends, work hours, and the timetabling of social events such as sports training or choir practice. Digital technologies disrupt social and clock time. They invite us to lose control over our personal time: for example, through always-on Wi-Fi at home, facilities to quickly purchase (impulse) goods in the supermarket, or everywhere-available access to work, colleagues, and clients. They discourage the development of a slow ethics of time. The technological prompt, it seems, is always to try and get more done in ever shorter amounts of time.

Clearly, a big part of the challenge of practising slow computing is to identify contingencies – moments in everyday life when you can refuse to play your part in hyper-communication and hyper-coordination without negatively impacting on your own wellbeing and ambitions and on the lives of others – and then deciding to participate in these social processes at a slower rate than might be otherwise expected. Today, for example, you might have read a book on the train, rather than answering emails. Ultimately, that move was about changing your use of time and space, although it probably did require some thought about how others might be affected: your colleagues might be waiting for a reply to an email and have grown accustomed to immediate responses during commuting times. But how often are your responses make or break? How often do they require a reply *right now*? Are these expectations universal, expected of those who walk, cycle, or drive to work? Is a more radical restructuring of routine daily life possible? Could you, for example, refuse to answer work emails or messages in the evenings or on weekends, or disconnect entirely on vacation? Or is some flexibility required, or even demanded, by others such as family and colleagues?

Our point here is that experiences of acceleration are often about the array of connections we have with others, each of them conceivably anxious about how we will reply. And they are also about expectations. Acceleration invites us to audit how we connect, yet also to ignore the possibility of stepping away and instead simply join in the intense flows of responses. There is always a mix of pressure and persistent temptation to remain connected. Many of us cannot escape acceleration – the 'slow' in slow computing simply isn't, or at least doesn't seem, an option for a large part of the day; this is just the way working life occurs today.

Nonetheless, we can, as it were, navigate or tweak it. When the pressure to remain connected is severe, for instance when work flows demand immediate responses from us, we can still make moves to slow down. We might wait until after breakfast before first checking email and notifications. By all means stay connected to email at lunch if your work culture demands it but turn off notifications from other messaging apps: the constant stream of WhatsApp messages from partners, friends, family, neighbours, and social groups draws you into chains of responses that amplify the pressure you already feel from an accelerating workplace. When back at the office, turn off your phone's internet connection and check your email less often: intense bursts of replies once an hour or even less often can allow your mind to concentrate on other tasks. You may feel the need to check in, but over time that sense of urgency can dissipate. Even when emails or other messages do arrive, you probably do not need to reply to all of them right now. Focusing on those messages that really *do* need your response can give you breathing space. Likewise, assess whether you really need to stay connected to social media while at work: Facebook or Twitter might be notifying you of news events, but do you need to receive these 'pings' all day long? And for that matter, all night long as well?

When the temptation is greater than the pressure, this is the time when we should *actively* strive to slow down. There are plenty of options. For example, keeping commitments to have device-free meals or exercise with others. The idea is to produce spaces and moments in which we can engage each other, without devices. Yes, 'devicing' is pervasive among adults and children: banning devices entirely is unrealistic for all but the most vigilant, committed, and in many ways privileged families (because devicing is, for so many of us, now a fundamental part of employment). Nevertheless, slow or slower moments can take shape if we turn non-devicing into a game. For example, there might be minor social penalties for the first person to check their phone at a dinner, such as doing the washing up.

More broadly, we can look to reserve computational-free leisure time by refusing to engage with colleagues outside of work hours. Setting up an automatic 'I'm disconnected' message on email can alert senders that there's little point in trying to contact you outside office hours. Casually informing colleagues that you're trying to minimize when you'll be connected might discourage them from getting in touch in the evening or at the weekend unless it's a genuine emergency. As part of this, why not switch off home Wi-Fi routers in the evening or at weekends and avoid the temptation to jump online and respond? Or, leave your phone in 'airplane mode' when you're trying to stay disconnected. And, as numerous commentators have emphasized, keep the phone away from your bedroom: its glare can prevent you from sleeping and, by offering numerous invitations to go online, it encourages you to speed up not slow down.[9] Viewing messages about issues and problems at a late hour is likely to set up a restless night's sleep. Sleep deprivation increases the chances of under-performance, making mistakes, overlooking crucial details, and reacting emotionally.[10] Your employer and colleagues might not realize it, but restructuring your time and rest in this way is almost certainly going to make you more innovative, productive, and pleasant to work with. In other words, it is to their advantage as well as yours.

Beyond changing how we use devices, there are other ways to slow things down and reduce the urge to hyper-coordinate. You can strive to organize pre-arranged meetings with friends or social groups, rather than trying to get together on the fly. Especially at work, but also in social life, there is a tendency to use services such as Doodle to find dates for meetings and the like, but this can demand so many responses that it works against the ethos of slow computing. A slower approach is to fix all dates at the beginning of a week or the year and then stick to them.

There might also be a case for reassessing our social circles. Do you really need to be reading about or interacting with 'six-degrees-of-friends',[11] many of whom you'd never heard of before social media? Perhaps you would be better off focusing solely on the one-degree friends that actually matter. Shrinking your social network might reduce your attention overload and eliminate your exposure to partisan Facebook posts or comments on Twitter that set your teeth on edge and heighten anxiety.

When out and about, too, we can resist the urge to speed up our lives. In the supermarket, why not wait in line rather than use self-scanning services (using the time to relax and mull something over)? At the train station, why not pick up and hold on to a paper timetable, rather than always using a smartphone app? Perhaps you could do your shopping in local stores rather than online. When you have ordered items online, why not collect those goods in-store, if possible, rather than have them delivered to your door? Simple steps like these, all of which undoubtedly reintroduce a degree of inconvenience, can reorient your life and help you reclaim a sense of ownership over your time and provide more 'authentic' and leisurely experiences.

A complementary approach is to assess how your device is arranged. Many smartphone users have the most common apps on their home screen for quick access, not least apps for email or social media. Removing them from the home screen introduces one additional step when you think about entering into those chains of responses; this can give you that split second to remember not to do so. Another option here is to switch off notifications or the 'push' option on email; this can give you some peace and quiet and compel you to actively receive updates, thereby giving you an extra element of control. An alternative is to have separate devices for home and work, making sure there are no work-related apps on the home phone, and only using the work phone during work hours. A more extreme approach, of course, is to delete social media or email apps altogether. You will still have the browser and so you can still 'enter', as it were, but that additional element of inconvenience can help you strike a better balance.

In short, pursuing a realistic sort of 'digital detox' is possible, albeit hard (in fact, it can be very hard for all sorts of reasons as we discuss later in the chapter). An extreme approach – getting rid of your smartphone or laptop – runs counter to the point of slow computing because it seemingly entails abandoning the joy of computing. As we emphasize throughout the book, there is a lot to like about digital life. The challenge is to strike the right balance. The key point is that technology can remain a part of daily life and some of the effects of social acceleration can be ameliorated if we decide to adjust or tweak our practices in ways that slow down our rate of participation. The 'slow' in slow computing is possible. We have options. At issue are everyday decisions about how we interact with the variety of technologies around us.

Toward data sovereignty

As we've just discussed, a major element of practising slow computing requires slowing down our interactions and reducing the number of times we take part in chains of responses. We may face constraints, but slowing down is an option we can all pursue to varying degrees. In contrast, negotiating data extraction is much more complicated. For one thing, data extraction can occur even when we think we're 'offline'. Companies and governments are collecting data about us in the supermarket or out on the street. They use these data reserves to develop new products or to alter facets of everyday life, such as traffic flows; we are in the scaffold of data extraction architectures whether we like it or not. Another complication arises from the opaque and, in a sense, passive manner in which data extraction now occurs. For example, a great many devices and services are tied into the largest companies (like Google, Facebook, and Amazon) without digital users knowing it. Fully tracing out the connections between your digital life and the universe of companies looking to make something of the data we produce is almost impossible. The data-laden, data-infused economy is vast. As digital theorist Katherine Hayles puts it, the 'ecology' we're highlighting here is 'too recursive and complex for any simple notions of control to obtain'.[12] There is too much taking place – much of it occurring offstage in private experiments or surveillance practices we have yet to learn about – to know how successful attempts to manage our data are in practice. Indeed, in Cheney-Lippold's view, the extent of data extraction infrastructures only leaves us a 'sliver of agency' to actively intervene and shape our data footprints and shadows. The task is to use that sliver of agency to 'mess' with data extraction and 'produce hiccups against algorithmic subjectification'.[13]

A guiding ethos for doing so is that digital subjects like you or us can work toward establishing a degree of 'data sovereignty'. Data sovereignty is the idea that we should retain (some) authority, power, and control over the data that relates to us, that we should also have a say in the mechanisms by which those data are extracted, and that other entities, such as companies and states, should recognize that sovereignty as legitimate.[14] The term has its roots in the claims of indigenous peoples to the right to maintain, control, and protect their cultural heritage, traditional knowledge, and territories, and determine and govern how data related to these are generated, analyzed, documented, owned, stored, shared, and used.[15] Such an assertion is made after centuries of data extraction made without prior and informed consent for ends that were to the detriment of indigenous peoples. This notion of data sovereignty has merit for all people.

Any data exchanges, such as the swapping of data for a service, should recognize and respect the rights and wishes of those providing the data. In other words, by staking a claim to data sovereignty we take back some control over how we produce data and what happens to these data subsequently. From a more collective perspective, data sovereignty proposes that principles and practices are developed for creating data commons rather than private reserves, or for data to be destroyed rather than hoarded and monetized. It is not to reject the utility and value of generating and extracting insight from data, but it seeks to shift who controls and benefits from this value towards those whom the data concerns. The challenge is to contest and transform how, and for whom, data extraction is working, without fundamentally changing or reducing our access to the types of digital services we enjoy. We want to hold on to the look and feel of, and the interactions we have with, computing while at the same time transforming the underlying logic and problematic nature of the emerging value chains.

None of this is easy. The present regime of data extraction is well embedded at this stage and our data sovereignty is only partially recognized. This recognition is, in many ways, reluctant and resistant, forced through legislation and regulation. Data sovereignty seemingly runs counter to the logics of capitalism and the free market, which treats us as consumers and data products from which capital can be accumulated and value leveraged rather than sovereign citizens.[16] It also runs counter to the governing principles of digital life: that we should embrace convenience, participate in chains of responses, and produce data in ways that expose us to the emerging constellation of effects. Nevertheless, we think working toward data sovereignty is worthwhile and viable, even for digital subjects who do not, and often cannot, develop a sufficient understanding of the full range of ways we are caught up in data extraction systems. For us, to claim our sovereignty through an assertion of our rights and by resisting present practices challenges companies and states to change their business models and practices. Such an approach requires us to continually strive to identify, consider, and adjust our practices, to learn how a territory and its terrain are changing, and to find ways to claim back some autonomy and contest any pernicious practices that are against our best interests. In what follows, therefore, we highlight some pathways through digital space that all of us can pursue. The task is to develop choreographies – 'data dancing' as we put it – that can bring us some sovereignty over our data. We highlight four key moves.

Curating your digital life

Data extraction architectures are pervasive and woven into the fabric of digital life. All of the devices and services we use today expose us to data extraction, with many having default settings that allow companies to monitor and record our various and numerous responses. Curating devices and services to reduce exposure is a relatively straightforward step that few of us pursue with enough vigour. It takes time and effort to do so, time we just don't seem to have in our accelerated lives. We use dozens of different digital devices, apps, and services at home, work, and elsewhere every day. That places a potentially large burden on anyone who really wants to curate their settings. Obstacles, however, are there to be overcome: constructing a degree of data sovereignty means we simply need to find the energy for, and make a commitment to, curation. For example, one simple step on either an Android or iOS smartphone is to reset or opt out of the 'ad ID'.[17] The general task is to become familiar with such settings and then to become accustomed to tweaking them, especially after software updates, when a militant curator who has kept track of these settings will check for sly changes that reopen the conduits connecting them to various data reserves. It might be the case that the settings we want to implement[18] will disable a core feature: in this case, we need to assess the trade-offs between service and privacy and decide whether we want to grant access. Sometimes apps change the terms of service, sometimes you change your mind as to what privacy settings you will tolerate. In either case, if you no longer want to accept the current settings it might mean having to uninstall the app altogether; if you have already been using the app, this will mean losing access to your own data history. Embracing slow computing might then mean giving up services you have come to enjoy. Data sovereignty has a minimal cost expressed in the time and effort to curate your settings, but sometimes greater costs when features of your digital life may need to be abandoned.

Beyond curating the settings, we can 'outsource' some of the work of data protection to privacy enhancement tools designed to limit the data harvested from our behaviour online, as well as enhance cybersecurity by protecting us from malicious sites: for example, browser plug-in tools such as 'Https Everywhere' (that always seeks to connect to sites using a secure and encrypted connection), Privacy Badger or Ghostery (that block trackers embedded into website code), and various ad and pop-up blockers. In addition, you should make sure you have virus and malware tools in place. In terms of hardware interventions, we can use USB 'data condoms' that are attached to USB power cables and prevent data exchange when your device is plugged into public ports, or camera covers that stop cameras capturing images if they are surreptitiously used by a third party.

More generally, if you are unsure of the provenance of a website, use an online reputation and internet safety service such as Web of Trust, Norton Safe Web, or Trend Micro Site Safety Center to see whether it is okay to visit or share data with.[19] Reputable sites will also display a trust certificate awarded by internet trust organizations, and sites can be counter-checked on the websites of these organizations.[20] They will also have privacy policies. You can also search for user reviews of websites and apps to see if anyone has had bad or good experiences. As a rule, only share data with sites that display a padlock and 'https' in the browser address bar and whose provenance you are sure of and feel you can trust. If a service asks for sensitive data outside of a task such as registration, login, or purchase checkout, then you should be cautious about sharing information.

As well as curating your devices and services, it is equally important to curate your digital footprints – the information you willingly share with social media and other apps. Here, curation is about reflexive and responsible sharing of personal information. It involves the active and ongoing consideration of whether we really want our thoughts, opinions, photos, videos, files, diary, and so on online as part of our reserves of data (shared via social media, email, cloud services, calendars). The only way to keep something truly private is to not share it via any digital device or service. Once something is online it is difficult to maintain control: posts can be cut and pasted, copied, shared, and recovered. Even if you immediately regret posting something to social media and delete it within seconds, someone could have taken a screenshot and recirculated it (as many politicians, celebrities, and ordinary people have found out to their cost) and it is still in the databases of the service provider. Anything you would not be comfortable sharing with strangers in a public place such as a high street, or with a newspaper publisher, you should think carefully about sharing. This responsibility does not just apply to our own data, but that of others as well. As a general rule, you should not tag people in posts or photos unless you ask permission or know that they will be okay with it. To be clear, we are not suggesting that you do not share any data – after all it is impossible not to leave digital traces – but that you should think carefully about what you do share and whether it is necessary or in your interest.

Part of this data curation is to actively manage how you use apps, the internet, and other devices. For example, you might swap to using a non-tracking search engine, such as DuckDuckGo rather than Google or Yahoo! Or you might start using private (incognito) browser sessions and to actively manage/delete cookies and browser history. You might limit the services you register with, only subscribing to those that you will use repeatedly. Or you might register for temporary services, like access to a store's Wi-Fi, using false details. Likewise, it might be an idea to opt out of unwanted mailing list subscriptions (especially those you were added to without permission) or sites storing your details, or to delete apps you are no longer using (unless you are confident you might use them again in the future). Don't do silly quizzes on Facebook or clickbait websites that are often designed to extract data used in security questions: 'Your porn star name is: the name of your first dog and your mother's maiden name'. Don't upload a photo, along with data about you, to a site whose provenance you don't know on the promise of getting an image back of you looking 20 years older. Now they have personal information about you – your face and name, the keys to facial recognition. For services like DNA testing websites, do you really want to share your DNA with a company that is going to monetize it in ways that you were not anticipating?

You might also consider limiting the use of Google, Facebook, or other platforms as the means to sign into other accounts, as this de facto enables tracking across sites. Likewise, while sharing your contacts, email address book, or friends/followers between platforms helps you connect with people you know, it also shares your social network with a new platform and any data brokers they are trading with. You could consider having separate work and home email accounts, so that personal information does not end up on work servers where they are open to discovery by the company.

In addition to curating active use of devices and services, it is possible to use data privacy and protection laws to request access to data that companies hold about you, to challenge the veracity of that data and demand correction if false, to see any decisions based on those data (such as a credit rating score), and also to request deletion and opt out of future data extraction. It can be difficult sometimes to know who has generated data in relation to you, or who data has been shared with. As such, the easiest places to start are the origin points and data brokers. You can access and manage, including deletion, all the data Google holds about you at: http://google.com/takeout. You can do the same for Facebook.[21] It can be more difficult to get access to data brokers, though legally you have entitlements to access data held about you. One data broker, Acxiom, has made a point of letting you see your own data, though it uses the opportunity to collect more data in order to help verify the veracity of its databank.[22] Within the EU, new laws enable you to pursue the 'right to be forgotten' and to have content from websites removed from search results.[23]

The core point here is that practising slow computing is about expressing your data sovereignty and becoming an active data curator where you take a more responsible role in deciding what data you choose to share. There are limitations, of course. Even when we do curate our devices or services, we cannot know if it will actually work. To take one example, the billions of phones running Android software designed by Google have numerous settings pages because there are so many apps operating within the device. We can tweak those settings as much as we want and try to ensure we're not exposed to Google's data extraction effort. But recent investigations have uncovered code within the Android software that overrides settings within individual apps, thereby allowing Google to acquire data about the location of each individual device.[24] There are inevitably many more cases that have not yet been revealed where other companies are using similar strategies to grab data about us. Likewise, state intelligence agencies are extracting data via backdoors and other tactics. Our participation in chains of responses has economic and intelligence value; fully stepping out of those chains while continuing to use digital devices and services is impossible. Even so, trying to curate, trying to express data sovereignty, remains worthwhile.

Using open source alternatives

One core element of the joy of computing is about accomplishing tasks; digital devices and services facilitate this for us to an extraordinary extent. We have numerous options when it comes to choosing which devices or services we use. Many of the biggest companies in this emerging ecology have successively used revenues to develop faster and smarter equipment and programs: their research and development budgets are enormous and the teams of engineers they put to work have produced incredible products. There is a mainstream ecology dominated by well-known names – Google, Apple, Amazon, Facebook, Microsoft, and so on – who market a range of products to billions of consumers around the world. Data extraction in one form or another is germane to what they do. As we have pointed out, this form of data–driven business model involves selling ads based on profiles we produce for them or enrolling us into platform services such as Microsoft's 365 suite of software products or Google's Android system, but this model is also intertwined with a push to develop new intellectual property. Slow computing and data sovereignty will involve all of us continuing to pass through their ecosystems at some point, whether at home or at work. The dominance of these mainstream players means we are unlikely to fully evade their data-grabbing activities.

Bound up with the mainstream, however, and at times offering a serious challenge to their dominance, is an alternative ecology constituted by a wide range of actors that open up pathways for us to step out of the mainstream. Open source alternatives are key because they do not have the commercial imperative to extract and monetize data. Rather they exist for the public good. Open source software are programs whose source code and design are publicly accessible and whose compiled code is freely shared. People can see and modify the program, and they can often use it for no cost (there may be a cost for a licence to use a compiled version of the code, though it is more likely that software services and supports will be charged for or a donation to the project costs requested). They are usually created through collective endeavours, with labour volunteered by enthusiasts to produce an alternative to proprietary software. Sometimes they are supported by foundations. The projects generally embrace principles of open exchange, collaborative participation, transparency, and community-orientated development.[25]

Consider the base camp for most of our computing: operating systems. On desktop computers and laptops, Microsoft and Apple are the dominant players. Both companies work hard to create impressive operating systems within which we can browse, play games, listen to music, and work. But there are other operating systems available and one group built using Linux, an open source code, deserves particular attention. Linux-based operating systems typically come with all the core functions we find in Microsoft and Apple platforms. The individual programs and apps are developed by teams of dispersed programmers and coders, many of whom contribute their time for free. Because their products are open source, their security flaws and other bugs are out in the open for other specialists to scrutinize and potentially fix. Strictly speaking, therefore, the transparency of open source operating systems should make them more secure, not least with regards to data extraction. Certainly, unlike a Windows-based operating system, none of the programs or apps should be secretly sending data to a parent company. And to the extent that the designers of programs learn from what users are doing when they develop new features, this form of data extraction does not result in a proprietary feature patented by Microsoft or Apple.

For these reasons, we suggest greater data sovereignty can be obtained when your slow computing takes place within an open source operating system.[26] One positive step toward slow computing would be to spend some time learning how to install such a system on a machine at home, for example. We expect you'll be surprised. Although the layout will be different, it will most likely be quite intuitive and heavily customizable, so you can rearrange things in a way you like (of course, this could take a bit of time to configure, and to maintain going forwards, but the investment will provide a system that fulfils personal preferences and data autonomy). Keeping in mind that Linux was made to address social needs, not profit, you might develop quite an attachment to it. The prompts and invitations that Microsoft has tended to incorporate into its operating systems, many of which raise the possibility of the company acquiring information about our computing habits, are nowhere to be seen. A Linux-based operating system is not one part of a larger corporate strategy unfolding within a global, near-trillion-dollar enterprise. There is no drive to learn how you swipe, tap, click, or scroll. In a basic sense, during the time you spend on a Linux system, you're left alone. Left to get on with accomplishing tasks. Stepping out of the mainstream and into this alternative ecology is a step worth taking if you want to create a more balanced digital life.

Desktops and laptops are not the only devices with alternative operating systems. Certain Android-based smartphones can be installed with a so-called 'Custom ROM', which replaces the operating system the phone came with, which is often tinkered with by the manufacturer or the mobile phone company. (If you own an iPhone, forget about it!) Installing a Custom ROM tends to be quite tricky; many of the newest Android phones are designed to make it incredibly difficult to do so. And the end result can be disappointing. There is a chance that some apps won't work perfectly. As such, we would only recommend trying to do this if you have a good degree of digital literacy and the time and confidence to tinker and fix computing devices. Even so, there is a lot to be said for giving this a try. For one thing, learning more about how your phone works is valuable. This is the one device we carry around so much and spend so much time staring at. Yet, unlike the desktop computer or laptop, we're discouraged and often prevented from fixing anything ourselves; few of us take any steps toward tinkering with the core system, even if many of us spend a fair bit of time altering the phone's overall settings or the settings within individual apps. If you do install a Custom ROM you'll have a phone based on the Android operating system but not necessarily one with all of the numerous Google apps pre-installed (adding this suite of apps tends to be an option left open for you to decide upon during installation). In essence, you could use a smartphone that does not feed data back to Google to anywhere near the same extent as your original device. It will still make calls, connect to the internet, take photos, and let you perform basic tasks. But you will have sovereignty over your data because you can decide not to accept the invitation to sign in to Google's data-grabbing architecture.

Beyond operating systems there is a diverse set of open source programs and apps, many of which provide high quality alternatives to proprietary software. For example, LibreOffice provides a set of alternative packages to Microsoft's Word, Excel, and Powerpoint. The Mozilla Foundation (slogan: 'Internet for people, not profit') provides an open source web browser (Firefox) and email client (Thunderbird). Whatever your interest and task there are various programs available. For example, if you are interested in creating and manipulating visual content there are numerous open source packages: for example, photo editing – GIMP; 3D modelling – Blender; vector art – Inkscape.[27] Our advice is to search for suitable programs and check out reviews to see which ones are best for what you are after.

Stepping out

If stepping into the world of alternative, open source operating systems and programs is one possible pathway toward slow computing, stepping out of the chains of responses built around data extraction is undoubtedly another. By stepping out, we mean leaving services and closing down accounts, or not using them to begin with. There are two key questions here. First, can you cope with the associated costs (to your social life or, depending on the platform, your position within your working life) of deleting your profile(s) on Facebook, Twitter, LinkedIn, Instagram, WhatsApp, Waze, Slack, Snap, Tinder, Uber, and so on, or shutting down file storing services such as Google Drive or Dropbox? Second, are you really going to gain sufficient value by upgrading formerly analogue devices to digital networked ones? Is it in your interest to have a networked fridge, coffee machine, or thermostat? To be sharing your TV viewing habits with the television manufacturer (along with whatever else you say in front of it) or to share all your questions and thoughts with Siri or Alexa? Is swapping data for a service in your interest?

Counteracting data extraction and pushing for greater data sovereignty could require that we abandon or forego digital services. The #DeleteFacebook movement, which sprang up in March 2018 amid revelations about Cambridge Analytica's use of data it gleaned from the social network, contributed to a dramatic fall in Facebook's market capitalization – and demonstrated a degree of interest among the population in changing how they interact online. Such moves tend to arise as a protest against the consequences of data extraction for privacy and political freedom. And for good reason. The extent to which Facebook's profits stem from the monetization of data on our responses – the extent to which we are the producers who make their platforms profitable – receives far less attention. Building a more balanced digital life, we argue, means reducing your exposure to data extraction for the sake of your privacy and because your compensation – your remuneration for producing – is negligible, at best. When we are active on Twitter or LinkedIn, Facebook or Uber, we add to the economic power of these companies and thereby endorse data extraction as a suitable business model. Data sovereignty is expressed when we reject their invitations and ignore their prompts.

A few years ago, the 'cloud' was a major buzzword in business circles. The buzz was created by companies such as Amazon, Microsoft, Google, and Dropbox promoting their ability to store personal or business data online, thereby removing the need for users to worry about their computers failing. Rather than taking photos and saving them locally, why not upload them to gigabytes of space on the cloud? Rather than a small business spending money on technical support to ensure a local server works well, why not simply pay for some space within the Amazon Web Services cloud? The buzz has dissipated but cloud services are now integral to digital life. Many of us now have documents, photos, and videos saved on the cloud and fret less about data stored locally. The advantage of this is familiar to anyone who uses multiple devices: desktop at work, say, as well as a laptop at home, plus a smartphone or two, and sometimes an iPad or tablet. Cloud services allow us to tie each of those devices to one place, which lets us access or add new files regardless of where we are (albeit, so long as we have internet access). Few developments over the last few years have presented us with such convenience. This sort of service exemplifies the joy of computing: easy, immediate access to piles of documents and albums of photos, with almost seamless new tools that enable us to share and even collaborate together online.

However, if slow computing is about trying to establish greater data sovereignty, relying on the cloud services offered by Silicon Valley giants or other commercial enterprises is far from ideal. Rather than add to the data reserve held by Dropbox or Google, which might expose you to ads they have sold or hidden (and conceivably, depending on the terms of service and whether the company respect those terms, illegal) analysis of your data, you could create your own cloud. Not easy. But not impossible, either. In fact, tying network-attached storage (NAS) to your home internet router does not require advanced technical proficiency. Setting up everything will take time and you might experience some difficulty in accessing your cloud from each of your devices. Still, with some effort, you can make it happen: your personal data will be accessible when you're out and about and you will have reduced your exposure to data extraction.

Of course, companies such as Dropbox or Google do not want any of us to do this, hence the offer of free space up to a certain limit and their development of new intellectual property to make their services run smoothly. Establishing data sovereignty means refusing these invitations. Slow computing entails embracing a degree of inconvenience and possibly accepting that we might need to give up some features of digital life. It does not mean your digital footprints will disappear from the prying eyes of government surveillance: employees of the National Security Agency, say, will find a way to access your home-run NAS cloud, should they wish to do so. In short, this move within the broader choreographies of data dancing does not magically guarantee complete privacy. But it does enable you to withdraw data from the commercial reserves; on that basis alone, building your own cloud has some merit.

Obfuscation

Our responses to invitations matter. A search engine invites us to find product reviews or information about a city we will visit. Our search reveals we are looking to buy a guitar or travel to Nigeria. Advertisers will pay to know this. Companies that track our responses can sell what they know about us to those advertisers. Their revenues can be used to develop new ways of tracking us. Obfuscation offers one way to make a mess of all this. Obfuscation is the deliberate use of ambiguous, confusing, or misleading information to interfere with data extraction. It could include tactics such as providing false answers, evasion, noncompliance, refusal, or even sabotage of the extraction process. As leading privacy experts Finn Brunton and Helen Nissenbaum note, one such form of sabotage is to create a fog of data – a proliferation which makes it difficult to spot the real data in a cloud of false readings;[28] it is a strategy to hide meaning within nonsense.[29] After installing a browser extension such as TrackMeNot,[30] for example, numerous search queries are made from within your browser, whether you are active or not. When you actually search, therefore, the companies tracking your activity cannot decipher your real search from the numerous other queries generated. Your actual response to the invitation to search is obscured, such that 'there's no reliable way to understand the difference between a genuine and a bogus query'.[31] AdNauseam[32] is a browser extension that automatically clicks every ad on a webpage making it impossible to work out which product a person might be interested in, thus disrupting the tracking of advertising networks. Another form of obfuscation is to use a pseudonym or a false identity, or several of them, to sign on to free Wi-Fi sites or register with services. Having different usernames and profiles across platforms makes it harder to identify data from a single user and join them together.

It is also possible to obfuscate by encrypting data and communications in order to evade data extraction. Your devices, apps within them, and communications via email and messaging can all be encrypted in ways that keep your data out of data extraction architectures (although possibly not government surveillance systems). Your browsing and searching can be encrypted. One possibility is to use a Virtual Private Network (VPN), which can be free but is usually a paid service. A VPN effectively creates a private network across the public internet by using authenticated remote access via dedicated servers and encrypted tunnelling protocols. While they do not make online connections fully anonymous, they can increase privacy and security.[33] An alternative is to use a Tor browser. This technology originates from within the US 'military–industrial complex' and is a source of controversy among technology experts because it remains unclear whether a government entity, such as the National Security Agency, can track your actions using Tor. Moreover, if you are the only user within a neighbourhood who browses the web using Tor, there is a heightened risk that your ISP will pay special attention to what you are doing as this method is used to access the so-called 'darknet' and used for illicit and illegal activity.[34] But for the purposes of slow computing, the advantage of browsing the web using Tor is that it disconnects your specific IP address – the precise location of your device, which identifies you or other users within your home – from your pathways through digital space. This is possible because Tor involves 'routing user traffic through three different medium servers [which means] the index of a single, unique user gets blown apart'.[35] Data about your moves through the web can be grabbed but the connection with you is obfuscated. You make a mess of data extraction systems because 'your metadata is just untethered, isolated nonsense'.[36]

Precisely because you are creating 'hiccups' when you use Tor, some websites won't work as you might expect them to. The web via Tor can be slower, less convenient, clunky. And with numerous other openings existing via which governments or companies can extract data about us – from programs that can record our keystrokes or apps that use our phone's microphone to listen to our conversations – even Tor users are unable to avoid data extraction entirely. As Cheney-Lippold puts it: 'No matter how hard we may try to avoid surveillance's embrace, its ubiquity still manages to bring us back into the fold.'[37] Nevertheless, acting on our 'sliver of agency' by obfuscating data and making a mess of things is an everyday form of resistance in the digital era. Such moves can sometimes entail embracing a degree of inconvenience, but this is the reality we confront today: we either accept that all of our actions online are a form of work for obscure companies that monetize how we respond, or we actively try to go on a 'data strike', which in turn means we need to go out of our own way.

Why practising slow computing is hard

All these tactics require effort and energy on an ongoing basis. Some require developing new skills. It is easy to lapse, or not even start. Indeed, there are a number of reasons why practising slow computing is hard.

First, there are social pressures to stay up-to-speed and in the (data-infused) loop. Digital channels have become the primary mode through which we communicate across distance with family and friends. That distance can be very small: we know of households where members message each other from different rooms. We schedule and manage our lives using these technologies. We have an expectation of others and ourselves that we will be always available, everywhere. We want to know that we can contact our kids instantly; that we can arrange a drink with a friend at short notice; that we can get timely advice from a sibling on the other side of the world; that we can interact with folks on social media. We want other people to be able to reach us if they need to do so. We are social animals, we like contact and exchange. We enjoy debate and competition. We like to feel involved. Not being online or available through a smartphone disconnects us from these networks.

Not checking posts or deleting an account is anxiety inducing: you could be missing an interesting or important exchange between friends, or the posting of an old photo from school you haven't seen in 15 years. There is strong social pressure then to keep practising fast computing. Any attempt to take up slow computing must try and reset social expectations and obligations and this can be difficult to do in practice and needs to be worked at over time.

Likewise, many employers and colleagues now have an expectation that you will constantly be available to answer emails, take a call, or get online and edit a document. That you can interrupt a romantic meal, step out of a concert, or interrupt a vacation to deal with a work matter. We often have that expectation of our co-workers, sending them emails in the evening and presuming we'll get a timely response. Fast, after all, is the logic of capitalism. There is pressure to become a workaholic. No surprises, then: trying to practise slow computing within this work culture and web of expectations can be difficult, especially when rewards such as promotion, salary rises, and bonuses are at stake. It is not impossible, though; even some scaling back might help you strike some balance. And it could even improve your productivity or the quality of your work, which your employer might recognize.

It is not simply social or work expectations that make us beholden to technology; there are also structural imperatives to be so. The state demands we interface with government through digitally mediated systems. Our employers insist that we use a suite of digital technologies to do our work. These systems and technologies force us to perform in certain ways and open us up to particular channels of data extraction. There are real systems of power at work here that compel us to act as desired and are difficult to resist without causing waves that can create hardship: we need the payments and services the state provides; we need the salary that a job provides (as well as its sociality and valorization). In addition, we are enrolled in technical systems, such as forms of surveillance and profiling, whether we want to be or not. These systems are outside our direct control and might be working in ways that are biased. It can be difficult to know they are being applied in practice, let alone know how to challenge their use. Trying to enact slow computing in the face of such state and corporate power can be hard, especially at a personal level. It is not impossible, however, especially when approached collectively, as we explore in the next chapter.

Another reason why slow computing can be difficult is more psychological. We might like and enjoy the fast pace of life. Some psychologists even suggest we are hardwired for it, getting a kick from the buzz and energy of doing things at pace (although it can also be exhausting).[38] We have certainly internalized the need for speed: that sense that time is getting away from us, that there isn't enough of it, that we need to save it where we can and maximize efficiency. Moreover, we have become used to it; we have become what Carl Honoré calls 'velocitized'.[39] That means we might be so accustomed to a fast pace that when we slow down it feels unnaturally leisurely (in the same way that dropping from 70 to 30 miles per hour while driving can create a sense that 30 is *way* too slow). We are easily bored and don't know what to do with ourselves without a task to accomplish. We have become attuned to being impatient. Waiting a few milliseconds longer than normal for a webpage to load is an annoyance. As Carrie Fisher observes, 'instant gratification takes too long'.[40] When we want something, we want it now. Slow computing means becoming patient, accepting some level of inconvenience, attuning our attention so that it's not constantly looking for distraction, and learning to appreciate a slower pace. However, this is not so simple in practice because many digital technologies are designed to be compulsive and addictive.

There is a reason that so many people check the apps on their smartphones dozens of times a day. We crave the hit of a new message, a comment, a new high score. We are addicted to the platforms that we use.[41] Nir Eyal, a consumer psychologist, argues that habit-forming, addictive technologies which compel people to repeatedly return to and overuse a platform all employ what he calls the 'hook cycle', consisting of four parts – trigger, action, reward, investment.[42]

The trigger is the hook, the proverbial itch that the platform scratches. It could be information, interaction, exchange, profit, or something else that if fulfilled will ignite an affective response. External triggers are designed to get you to the platform and could include paid advertising, email prompts, PR, memes, and various forms of clickbait, such as silly stories or rewards. They tantalize us with the suggestion that visiting the site will offer some kind of reward (such as entertainment, amusement, friendship, free stuff). Internal triggers are conscious or unconscious thoughts that prompt us to return to a platform to rescratch a persistent itch. We already know what to expect from the app and we want more. It's our internal trigger that makes us check social media 50 times a day. The strongest triggers are typically experienced when we feel bored, lonely, confused, fearful, lost, or indecisive, as we are more receptive to stimuli and the app provides an instantaneous response and (temporary) relief to these feelings.

The trigger induces you to progress to stage two: action. You act on the trigger and start to use the platform – scrolling, clicking, typing, uploading. In addictive apps actions are streamlined, delightful, and as simple as possible to perform, eliminating as many steps as possible from the required process. Moreover, the action is achievable without conscious thought and therefore habit-forming. We can quickly scroll through a timeline, click a 'like' button, add an emoji, press a share button, see comments and replies, and so on. To deliver on the promise of the trigger, the action needs to lead to stage three: the reward.

The reward might be something entertaining (for example, a meme), useful (information), stimulating (a response to a social media post or a prize for a top score), or profitable (a discount or change of user status). For the reward to be habit-forming it needs to be variable, thus adding a sense of uncertainty and anticipation. If the reward is always the same, then the action will have a diminishing affective pay-off and you will eventually lose interest. In social media, the variable reward is finding out who and how many people reacted, both positively and negatively, to a post, photo, or comment. In a game, it might be points scored, bonus points, or access to hidden levels or weapons. From a neurological perspective, both the anticipation of a reward and the reward itself produce a spike in dopamine: that is, they create a chemical feeling of pleasure in the brain.[43] It is this activation of dopamine that makes the cycle addictive.

The action and reward then translates into the final stage: investment. Investment ideally involves moving from simply browsing a platform to becoming an active contributor, helping to provide content. People tend to value things they have helped create far more than something simply produced for them. By investing in a platform, creating their own profile and adding content, they have a stake in its success, in itself forming a trigger for seeing how their investment is performing – and hence looping back to the first stage. Investment is important for the entire tech industry – hardware and software companies – as it provides the basis for revenue, either directly through payment or indirectly through advertising revenues and deals with data brokers. For example, for the companies to make income and a profit, investment should ideally be more than time, labour, and content, such as a subscription to a premier service, buying in-game upgrades, purchasing next-generation devices, or clicking on an advert.

This four-part cycle has proven to be a very powerful promoter of digital platforms and helps explain why millions of people are seemingly hooked on constantly checking their emails and social media, playing computer games, and performing online gambling and shopping.[44] In the latter case, rather than trading data for virtual rewards, there are potential real financial penalties as people gamble and purchase more than they can afford, ending up in debt with all the associated consequences that entails (family hardship, losing possessions and home). The cycle is why many digital companies have rapidly become multibillion dollar enterprises.

Companies are well aware of the power of this cycle, which is why many of them employ neurologists and behaviourists to try and maintain their freshness (so that users won't get bored and move to a rival product) or make their products more compulsive. These researchers seek to refine the hook cycle through continual A/B testing, trialling new triggers, actions, rewards, and investments. Well-developed platform ecosystems, such as Google, Facebook, or Twitter, often have dozens of ongoing trials operating at any one time and they have a ready army of testers: us. We act as the guinea pigs, interacting with interfaces that are constantly being tweaked to keep us interested and addicted to the service, or willing to provide investment. Indeed, trials can involve hundreds of thousands of users.

As we see it, therefore, slow computing needs to be about realizing that these platforms are psychologically compelling, habit-forming, and addictive. They prompt us to act in ways that we might not be initially anticipating (such as, getting into a heated debate with a stranger, making an impulse buy, signing up to a subscription).

To intervene, we might not need to go 'cold turkey' or delete all our accounts, but a more balanced digital life must surely mean we should manage interactions more thoughtfully. To use an analogy with imbibing alcohol, we're suggesting drinking in moderation rather than to excess: to think clearly about what you are taking on board, and why.

Even if we want to try and unshackle ourselves from certain platforms yet still experience the joys of computing, it can be difficult to achieve in practice because we are so beholden to them; they are so thoroughly integrated into our digital lives. In 2018, journalist Daniel Oberhaus tried to live a month unplugged from Amazon, Apple, Facebook, Google, and Microsoft technologies and platforms – what he terms the 'big five tech companies'.[45] He found it relatively straightforward to give up Facebook and Apple, not missing his social network and not being an Apple product user. As a long-time user of Microsoft's operating systems and products he found the transition to Linux and open source alternatives a bit of a trial but managed it with a bit of persistence (which would no doubt also be the case if he was an Apple user). The other two were more difficult. While he could shop online elsewhere, a very large number of websites and services are hosted by Amazon's cloud. He found Google impossible to escape, in part because he uses so many of their services and his phone uses Android, and in part because his employer is also dependent on Google. Six months after his experiment, Oberhaus had drifted back to using Microsoft's operating system, had resumed some purchases with Amazon, and was still thoroughly entangled with Google's ecosystem. The stickiness of these companies helps to explain why their collective market valuation was worth over $3 trillion in 2018.[46] However, he had managed to alter his computing practices, in particular using open source alternatives that reduced some of his data footprints and shadow.

Finally, it can be difficult to imagine an alternative kind of life and society to the one we presently live in. While you might appreciate that the slow computing tactics we are promoting might help you personally to some degree, it's more challenging to believe that they will change the fundamental nature of society. They appear more like coping strategies than critical interventions, and while they might make some difference, this will be relatively minor in countering prevailing forces. Mark Fisher, in his account of why capitalism is so resilient as a political economy, argues that it has created a 'pervasive atmosphere' that is so thoroughly interwoven with our thoughts and actions that it is seen as the only viable system – the natural order – so much so that it is difficult to imagine, let alone practise, alternatives to it.[47] Lina Dencik uses Fisher's ideas to argue that surveillance capitalism is now so thoroughly embedded into society that it too has become accepted as inevitable and, more troubling, as commonsense.[48] In this sense, it is hegemonic: that is, thoroughly institutionalized and normalized so that it operates without having to resort to coercion. It is the accepted way of things.

In other words, a 'data doxa' can be said to exist. Gavin Smith explains that a 'data doxa' is a situation in which the present configuration and workings of our data-driven society and economy have been normalized as necessary and legitimate, making them difficult to challenge.[49] Smith argues that three types of data-based relations contribute to the formation of this doxic sensibility, mediating how we come to understand, accept, and internalize their logic, and how they become accepted as taken for granted, the way of things. First, is fetishization: data have been valorized extensively as *the* solution to a myriad of societal problems and the key driver of economic growth for the 21st century. How can one be against social and economic progress, which surely is in all our best interests? Second, is habituation: digital devices/data have become so thoroughly woven into our lives that it becomes difficult to imagine life without them. Third, is enchantment: we are so seduced by the power of digital technologies to enhance and enliven aspects of our everyday lives that we are prepared to live with any negative aspects. Together, these data-based relations make us desire and accept the present data-driven orthodoxy, regardless of any downsides, and in so doing reproduce the status quo. Indeed, they work to make it difficult to imagine or yearn for alternatives.

Pushing back against acceleration and extraction can seem like pushing a heavy ball up a hill, or acting like King Canute trying to hold back the tide. It induces a feeling of 'why bother' if ultimately you are going to succumb to the natural order of things? The data doxa works to disempower and discourage people from seeking alternatives. Importantly, through these examples, Fisher, Dencik, and Smith all make the same point: there is *nothing inevitable* about how society is organized and run. Indeed, the first point of resistance and change is to recognize the contingency of our political economy – it is open to being reimagined and remade. Fast lives and excessive data extraction are not the natural order of things. We can push back against them individually, using our agency to try and carve out some respite and different ways of living, and we can work collectively to transform society through social, cultural, governmental, regulatory, or legal means. Practising slow computing is therefore one component of enacting an emancipatory politics that can 'make what was previously deemed impossible seem attainable'.[50]

In short, slow computing needs to be worked at on an ongoing basis. It will be hard at times, but also rewarding. Occasionally you will feel like King Canute trying to hold back an inevitable, speeding, grabbing tide, and it will be difficult to try and hold the line. But the key thing to remember is that practising slow computing, even partially, should have a positive effect on your wellbeing, and on the wellbeing of others. It is possible to imagine and make individual and collective changes.

A slow computing day

Without any question, a slow computing day entails computing differently. It requires that you adjust your practices in lots of ways, many of which you are already doing, at least to some extent. Many people are already stepping away from the headlong rush into social media and smartphones and constant pings of emails or messages. Resisting the invitation to participate in the chains of responses that digital life places in front of us can help us reclaim our time and even some of the privacy we have lost amidst a data–infused world. As we hear more about the emerging form of digital society, more and more people are, in effect, tracing out data choreographies and doing some data dancing. But how should we go about working out what dances to perform?

Back in 2005, one of us co-authored two essays about the effects of digital technologies on our everyday lives. The first paper detailed day-in-the-life accounts of three individuals of different ages and backgrounds, documenting all the various digital technologies they encountered from when they woke up to when they returned to bed.[51] The second paper used the contents of a wallet, and all the unique codes on receipts, driving licences, credit and loyalty cards, money, and so on to examine the ways in which these codes are used to enrol and capture us, the objects we use, and the interactions and transactions we perform in data-driven systems.[52] Both of these essays used an audit approach to uncover and think through how digital technologies were reshaping how we experienced the world at that point.

In a similar guise, we think conducting such an audit might be a useful first step in practising slow computing. There are three steps. Step one is to document all the digital technologies you use and encounter in everyday life across all domains – home, work, the street, travel, consumption, play. They might be ones you directly choose to use that create digital footprints, such as smartphones or smart televisions, or ones you have no choice in encountering that produce digital shadows, such as surveillance systems. Remember, some of the technology is backgrounded or hidden, such as smart meters, modern cars (effectively a set of computers on wheels), or sensor networks that track a phone's location. For smartphones and computers, detail all the specific apps and software you use. We suggest doing the exercise as diary entries for a single, typical week, as that length of time will capture all the things you do routinely and those that are more occasional. The diary will also give you a sense of how much time is expended on each activity and encounter over the course of a day and week. You can either do it as a proper diary, detailing your digital encounters as you go along or compiling it at the end of a day, or you can do it as a mental exercise, documenting typical days in a week from memory. Either way, it should not take you long to do, and if you can't find the time to do it, then it does suggest that you do need to embrace slow computing!

Once you have conducted your audit, you need to take the second step towards slow computing. This exercise is more reflective and considers the effects of these technologies on your everyday life. Divide up a sheet of paper into five columns. In the first column, list each of the digital technologies encountered, ordering them by those used and encountered more frequently. In the second column, detail how each technology affects the temporal organization of your day in positive and negative ways (for example, saving time, creating efficiency, enabling multitasking, lack of family time, feeling harried and stressed). You might also want to consider the overall benefits or costs more generally (for example, being more productive, or the effects on your enjoyment or health). In the third column, set out what this might look like if a slow computing perspective was employed: what would it ideally be like in terms of time spent and outcomes? In the fourth column, set out what data are being generated by that digital technology and what you think those data might be used for. In the fifth column, jot down how this might change in a slow computing day. Our discussion in this chapter provides some suggested tactics. You should find that some digital technologies are relatively benign in acceleration and extraction terms, some create more issues due to their acceleration qualities, others due to the data extraction, and some are problematic on both fronts. Some have paradoxical qualities, being enjoyable or productive but also creating over-burdening and stress. And some are easier to address than others.

The third step is to consider how you are going to adopt a slow computing lifestyle and to plan how it will change your existing diary and data practices. We think a useful exercise here is to use the second exercise, especially columns three and five, as a foil to plan an ideal slow computing day on a workday, on the weekend, and on vacation. You'll need another sheet of paper, dividing it into two columns. The first column is your proposed timetable for each day; this is your plan to deal with acceleration. The second column is how you propose to interact with each technology, or your proposed replacement technology; these are your proposed solutions for extraction. We are not suggesting that these become rigid templates for you to follow religiously. Life is far too fluid and unpredictable for that, and variety and serendipity are healthy and desirable. Rather, the proposed plan and solutions provide a touchstone for taking back control and creating a balance that is hopefully manageable.

These exercises are designed to provide a route towards a balanced digital life. We think they'll help you encounter new openings and opportunities. They are about your digital life and ask you to reassess what you can do, on your own, at your pace, and with respect to the social relations and networks that matter in your life.

But although having a slow computing day is definitely about us making new decisions on our own, it also involves other people.

For one thing, slow computing means looking to make the most of technologies developed by others and then installing them on our devices or incorporating them in our digital lives. We've highlighted a few of these technologies. More are taking shape. Our ability to pursue slow computing hinges on further developments that react to the changing face of digital society and digital lives. Staying alert to these developments can pay off. For every new feature or device that tries to suck us into acceleration and data extraction, there will usually be some counterfeature or 'plug-in' that we can use to protect our ability to control how we experience the joy of computing. For these reasons, we think slow computing stands a greater chance of success when we work in groups, workplaces, and institutions to alter how we co-exist in digital space. There is a collective element driving the seemingly always-on invitation to participate in chains of responses. Individuals always exist and act in relation to wider structures of constraints and affordances. It's important to talk up what you or we can do. Social change certainly requires individuals to act and produce change on their terms. But emphatically we also cannot ignore the broader scene: societal developments, the locations of individuals with respect to more general practices, actions, and ideas. In short, having a slow computing day – creating a slow computing life – pushes all of us to work collectively to remake digital society. It is, therefore, to collective efforts that we must now turn.

5

Slow Computing Collectively

Friday. 5pm. The strangest thing just happened. An email from management arrived saying: 'In accordance with the firm's new work-life balance strategy, the email client will be paused today at 5.30pm and will reopen on Monday at 8.30am. All messages sent during the weekend will be held by the client and delivered on Monday morning. All staff are requested to enjoy the weekend. Management are exploring the possibility of also pausing the email system overnight during the working week.' You don't get it. What's got into them? A colleague stops by. Apparently, the union pushed for this. Management agreed because they thought it'd be good PR. The local news channel's running a story on it tonight. Whatever, it seems like brilliant news and the office is buzzing.

It's hard to believe that any company would make such a move, right? Or that an employer would install an open source operating system on all of its computers and encourage all staff to use a Tor browser. But why not imagine such collective moves toward slow computing? For all that slowing down requires individuals to adjust their practices and create individual choreographies of data dancing to secure data sovereignty, slow computing *collectively* is ultimately what's needed. Workplaces are as good an arena to begin as any. In fact, given the reality of 'working time drift' there are few sites as appropriate to implementing the slowness in slow computing. In this regard, therefore, we tip our hats to the French 'right to disconnect' policy, a move akin to what we have mentioned in our imagined example. Workers in French companies with more than 50 employees now have the right to negotiate times when they will not be obliged to check emails or text messages.[1] Companies in other jurisdictions have implemented similar policies aimed at protecting their workers from stress-related illnesses and burnout and ensuring they get suitable rest. For example, Volkswagen blocks work email being sent to the mobile phones of workers between 6pm and 7am, and Daimler permits workers going on holiday to automatically delete all new emails while they are away.[2] Those seem like positive steps in the right direction.

As we want to demonstrate in this chapter, so much more is possible if we pursue slow computing collectively and on multiple fronts: industry, government, civil society, non-governmental organizations. It is an opportune time to dwell on these possibilities. More and more people are becoming aware of the extent to which they are tied into digital chains of responses and subjected to data extraction. Problems emerging in the digital society taking shape around us are becoming apparent, with the tensions inherent in digital societies coming to light like never before. It's time to push on from making individual adjustments to consider broader (possibly even society-wide) moves that celebrate the joy of computing while exerting new controls over what our love of 'devicing' should produce.

The rest of this chapter focuses on some of the practical dynamics of slow computing collectively. We first examine two connected components of collective moves to slow things down: slow computing practices and rights, on the one hand, and the creation of slow computing spaces, on the other. We focus on some workplace issues but also make numerous nods to opportunities for collective moves in other arenas of life. We think many of the possibilities we raise are well within our grasp if we push for them collectively.

Next, we consider collective action to evade data extraction. There is a lot of material to highlight here. We begin by looking at how industry might change to take the lead in addressing the negative side of data extraction through self-regulation. We then consider the role that political parties and government can play by driving policy and regulation that protects citizens. Finally, we examine the ways in which communities, civil society, and non-governmental bodies can foster data sovereignty in practical and political ways, including placing pressure on companies and governments to address the pernicious effects of data extraction.

In the third and final part of the chapter, we raise some issues about what a collective push to embrace slow computing could mean for the world at large. Just what would a slow computing world look like and is it a world we really want to live in? For all their faults, digital technologies create new and exciting possibilities. Is creating a slow computing world worth it?

Slow it down *together*

The 'slow' in slow computing is once again at issue here. Our call is to reflect collectively on how we participate in acceleration and explore ways of working together to adjust practices and slow things down. The challenge is to identify specific ways we can regain some control over digital society. What's required, we suggest, is developing slow computing practices and rights for entire groups of people, such as workers, and the creation of slow computing spaces in which one can step in and out of the fast lane. Our focus is on the workplace as a frontline site where collective moves are possible and needed. However, it should be noted that such ideas can also be practised in relation to family, friends, and civic society.

Slow computing practices and rights

Developing slow computing practices as a collective move requires a serious shift in thinking and general practices within workplaces. Rather than heightening expectations around work hours, valorizing the exploitation of labour, and pursuing fast computing, attention is shifted to quality of work, workplace satisfaction, and worker health, recognizing that these will improve worker commitment and productivity, and reduce lost days and staff turnover. Rather than submit to the temptation to accelerate by adopting new software or automated work flows, or rolling out devices that encourage working time drift, managers and workers can explore ways of embracing slow computing principles in their everyday practices. This is not only about disconnecting or unplugging – it is more than a matter of 'digital detox' – because slow computing is still centred on computing and enjoying facets of digital life. The question is how to create slowness within proliferating chains of rapid responses; how to identify where and when collective moves to slow down

111

will give workers space to breathe without necessarily impacting negatively on others, the broader ecology of the workplace, and the bottom line.

Auditing where and when groups engage in accelerating practices is a first step toward identifying what might be done to slow things down. When information flows are centralized, for example, there is a case to be made for restricting the periods when general information is sent: rather than emails or messages bouncing around all day long, whenever information is ready to be shared, central administrators could hold on to them and send everything early the next day. Even when information flows are decentralized, organizations can encourage team members to send only urgent messages during the day and hold non-urgent information for the 'morning news', as it were.

A further advance might lead toward slow computing rights. Something similar to what we have in mind is already taking shape, as we mentioned at the outset of this chapter. The right to disconnect, as posited by French and Italian lawmakers and some prominent companies (most notably large German enterprises such as Allianz, Bayer, Daimler, Henkel, Telekom, and Volkswagen), gives employees the right to stay offline outside regular working hours.[3] It is a relatively significant development against a much broader shift toward always-on communications, though there are caveats and loopholes, and penalties are light. Its adoption points toward a growing awareness that remaining tethered to a phone or other device and participating in chains of responses all day long can be harmful for employees, and thus the company. In the context of neoliberal mentalities about personal responsibility and self-governance – mentalities that, at a minimum, attack notions of solidarity among workers and, at their most vicious, justify anti-union legislation – the argument that workers deserve the right to disconnect is more radical than many critics might, with quite a bit of justification, point out.

The notion of having a right to disconnect might need to be extended so that workers have broader slow computing rights. Consider here the matter of neurodiversity,[4] specifically the possibility that acceleration is experienced differently within a workforce and the likelihood, therefore, that some workers are heavily disadvantaged by non-stop pings, beeps, requests, and the need to offer responses. Styles of both learning and working differ, yet the drift toward Slack messaging or project management in environments such as Microsoft Teams neglects and seeks to override difference, on the assumption that everyone can and should participate at the same pace and level of intensity. The call for slow computing rights is about recognizing that many of the technologies adopted within workplaces go too far: the right to disconnect outside working hours is, on its own, a necessary but insufficient move.

Given that many companies will want to maintain the status quo, indeed might want to further undermine workers' rights with respect to working hours and work conditions, gaining these rights is likely to mean moving beyond voluntary moves by owners and managers. Instead, it will require negotiated agreements about where the limits of acceleration reside. In other words, there will need to be collective bargaining of working hours, worktime organization, and the use of the 'digital leash' by companies, either through a labour union or employee negotiation within individual firms and across sectors or through government regulation and legislation. At present, working hour rights vary across jurisdictions. Across Europe, for example, there are four working hour regimes: pure mandated (set by the state); adjusted mandated (set by state mandate, with some adjustment through negotiation with sector and companies or with companies and individuals); negotiated (weak state mandate and primarily driven by sectoral agreements, with some limited negotiation with companies and individuals); and unilateral (negotiated between sectors, companies and individuals).[5] Clearly the scope to achieve slow computing rights across these jurisdictions varies with context, but nonetheless the ambition should be on the agenda of future revisions.

Slow computing spaces

We suggest that collective moves to slow things down could also involve agreements about the virtue of having slow computing spaces where workers can share ideas, write, plan projects, or conduct analyses, or even just relax and have a coffee or lunch without having access to the internet. The trend over the last 20 years has been to extend Wi-Fi everywhere. Most of us are glad of this and we look about in frustration when we cannot find Wi-Fi in a public space. Wi-Fi technology has been normalized; we expect it wherever we go. Imagine trying to push for its coverage to be reduced! But if slow computing is something to be embraced collectively, perhaps such a move is necessary. In an office canteen, a foyer, or some other zone where workers congregate, slow computing as a collective move could mean switching off the Wi-Fi router and putting in its place signage to indicate that this zone is supposed to be a slow computing zone where connection to the internet is discouraged, and where users are encouraged to take structured rest and recharge, or to work without distraction.

Of course, many users will simply switch over to their phone's mobile internet connection unless new norms are established that frown upon such a move. Such a zone could be akin to the quiet carriages on some of today's trains. As should be clear, however, making such a zone a reality is never going to be easy. It requires someone or some group deciding to push against normalized practices, communicating their concern and idea to a decision maker or committee, and then via the means of signage or everyday practice trying to communicate the point of creating a space where devicing is discouraged.

A more aggressive move, therefore, is to embed slow computing within the architecture of a building, as has been proposed by the Dutch architect Rem Koolhaas;[6] by Space Caviar, an Italian architectural practice calling for architecture that can include a 'space of selective electromagnetic autonomy';[7] and as demonstrated by the On/Off box designed by Sibling that creates a Faraday cage around a workspace which blocks all signals.[8] Such architectural moves are taking shape in some homes in San Francisco by 'applying heavy graphite-based paints on all interior surfaces. Conductive wire tape was then applied in a large network, connecting the walls, floors and ceilings. Finally, the windows were coated with transparent, EMF-resistant films.'[9] Entirely shutting down connectivity does seem extreme, but collective efforts to slow things down may require dramatic interventions. When companies or institutions develop new buildings, perhaps agreements about creating slow computing spaces are called for: rather than building connectivity into every device or appliance, as in the internet of things model, a more astute move may be to design spaces where slow computing is facilitated. The point is to create spaces where the temptation, if not the pressure, to connect is shut down.

From individual to collective data sovereignty

Challenging data extraction architectures means building greater levels of data sovereignty both at an individual level (see Chapter 4) and, as we now discuss here, via collective efforts. Beyond the choreographies of 'data dancing', there is an urgent need to identify and dwell upon broader moves that can produce 'hiccups' in the system and engender slow computing. But in doing so, it is necessary to bear in mind that the issues here are complex and involve a wide range of actors and stakeholders, including: billions of individual users with their numerous accounts and profiles distributed across a vast array of services; corporations who create devices and provide the services that permit them to gain from harvesting data; government regulators who claim they want to exert some control over what's taking place; influential organizations in civil society that monitor and challenge actions in this sphere; and social movements protesting dataveillance. Together, via various iterations and engagements regarding diverse issues, these actors and stakeholders contest the nature of today's digital society and economy.

There are, as we will outline, some impressive ideas and practices that yield collective moves to undermine data extraction and strengthen this axis of slow computing. There are signs that the nature of today's digital society could be transformed in productive and positive ways in the future. But there are numerous strong forces at work that present serious obstacles in the path to slow computing. For all the hype and PR – the promises not to be evil; the pledges that companies and governments will protect our privacy – digital society is driven forward by business models and governance structures based on data extraction. Collective efforts to create data sovereignty need to work with tech companies (not just the big five of Amazon, Apple, Facebook, Google, and Microsoft, but the whole ecology from start-ups to major players), but must also recognize what is truly at stake for these businesses. Our purpose here is to explore ways to collectively work with and challenge data extraction so as to create a data landscape more attuned to slow computing principles. We divide our discussion into three sections to consider: (i) what tech firms can do; (ii) the role government, policy, regulation, and law need to play, and (iii) the role of community, civic society, and non-governmental bodies. In our view, none of these can solve data extraction concerns on their own: all three need to be active and working in concert for us to make real progress towards a slow computing future.

We are not looking to halt or dismantle a data-driven society or economy, or to implement heavy-handed or expensive solutions. Rather than being unnecessarily restrictive, we envisage trying to formulate balanced responses, mutual consensus, collaboration, and win-win situations. At the same time, we recognize that interventions are needed to curtail some practices; some interventions might need to be regulatory in nature. We cannot be expected, as Julie Angwin notes, to simply 'hand over all our data and "get over it"'.[10] Companies and states need to recognize the real concerns of people and to act responsibly and for the public good. Our focus, then, is on pragmatic, practical, and political interventions.

The role of industry

Companies are clearly key actors driving data extraction. If they were to change their ethos and practices with regards to data capture and privacy, introduce market-based solutions and self-regulation, and tweak their business models, it would have a profound effect. Digital industries would prefer self-regulation rather than restrictive, regulatory interventions. They argue that formal regulation imposes unnecessary bureaucracy and financial overheads and that, on the one hand, the market is capable of acting responsibly and, on the other, it naturally self-regulates for fear of losing customers and market share.[11] There are several problems with this argument. First, it assumes that individuals are aware of and understand any abuses that might be happening, even though these take place largely out of sight, and that they have the freedom and choice to move their custom if they are unhappy. In reality, some companies are effectively quasi-monopolies, with consumers having few other choices for equivalent services. Second, while some companies will be ethical and conscientious in seeking to ensure data privacy, market regulation does not solve the issue of vendors who wilfully abuse privacy rights or are negligent in their data security practices. Third, the evidence suggests that companies generally only change privacy policies to favour their own interests or when under duress and to comply with legislation and regulation.[12] Indeed, in many cases, privacy policy changes update terms and conditions to cover more extensive data generation and data usage and to further limit the company's liabilities. Fourth, many companies actively lobby to reduce data protection provisions designed to address many of the issues detailed in Chapter 3, rather than embracing what society desires.

Lastly, there is lots of evidence that tech companies largely ignore repeated negative publicity concerning their data extraction activities, despite there being an awareness inside a company that the construction of reserves of data to sell ads and develop intellectual property has gone too far. A good example of such awareness is illustrated by the experiences of Sandy Parakilas, who was an operations manager in Facebook between 2011 and 2012. He described finding himself as working for 'a company that prioritized data collection from its users over protecting them from abuse'.[13] Crucially, the problem extended beyond Facebook because app developers in other tech companies were able to collect reams of data without violating Facebook's rules: 'Once data passed from the platform to a developer, Facebook had no view of the data or control over it.' As the *New York Times* has exposed, Facebook also developed 'data-sharing partnerships with at least 60 device makers – including Apple, Amazon, BlackBerry, Microsoft and Samsung', which enabled numerous tech companies to collect and analyze user data.[14] For Parakilas, then, Facebook's power to access personal data and its tacit approval of sharing data with other companies produced a 'dangerous mix' and a need for greater regulation. He quit working for Facebook in 2012.

Around the same time, Tristan Harris, a Google product manager, also reacted negatively to what he was seeing. He created and shared a presentation with other employees to raise the alarm that product developments aiming to encourage addiction were problematic. The presentation specifically criticized technologies intended to create distraction and 'called for Google to reduce the frequency of interruptions, batch notifications into a digest, and insert friction into the process of checking the phone to encourage people to use it less'.[15] Harris went on to found the Center for Humane Technology, whose aim is to 'realign technology with humanity's best interests', with support from some well-known keynote advisors, such as Robert Lustig, Tim Wu, and Cathy O'Neil.[16] The Center reflects an awareness by some within Silicon Valley of a need to promote a more balanced digital life. It calls for a new ethics with designs that undermine the drift toward technologies that encourage addiction, target children, and challenge democratic norms. It also calls on individuals to 'take control', for example by removing social media from phones, charging devices outside the bedroom, or reducing the number of apps on a home screen. In essence, the call is for individuals to pursue a degree of slow computing.

Beyond the substantive issues that these examples raise, they also highlight that those who work for tech companies have some agency to challenge from within the company vision, mission, values, approach, business model, strategic partnerships, and product design. Workers have the opportunity to influence in-house debates and to steer how the sub-unit they work for, and thus how the company as a whole, operates. Admittedly, this influence might be small at the individual level, but if many workers start to say similar things or work collectively then they can influence the company as a whole. The task is for those working in the tech industry to push their industries to reshape their underlying ethics and practices. After all, when we consider what sort of technologies these companies are working toward, it strikes us that we are still at an early stage in the overall game. There is significant scope to change the path along which they are developing and how they might operate in the future. There is no inevitability that excessive data extraction will remain a core feature over the long term. Indeed, slow computing could be the next 'killer app'.

Publicly, Facebook and Google, as well as other tech giants like Apple, have suggested their latest technologies respond to these types of complaints and concerns. Mark Zuckerberg, Facebook's CEO, has pledged to strengthen privacy and security processes. Apple has pitched Screen Time as a feature in iOS 12 (released in 2018), which it claims will help users control their use of devices. And with its launch of Android 9 Pie in 2018, Google announced a response to addiction concerns by releasing information on 'digital wellbeing' tools, which include ways for users to reduce the number of interruptions and notifications they receive.[17] On face value, there is a lot to like about all this. These developments are akin to the type of 'humane technology' Harris advocates. They reflect moves to create something conversant with slow computing. But at the same time, we must chuckle at the irony of Google encouraging users to use their phones, or Google's Smart Assistant, to act as the antidote for online engagement. Moreover, the company's entire ecosystem is based on data extraction to sell ads and develop intellectual property. Admirable as they may be, efforts such as Google's digital wellbeing strategy appear to be minor features in an otherwise densely populated data extraction landscape.

Rather than self-regulation, then, it might be competition that drives companies towards slow computing. Some enterprises see user privacy and data security as a means to achieve competitive advantage, developing privacy and security protocols and tools that will attract consumers away from other vendors.[18] For example, companies are developing products that have limited tracking or profiling, or employ end-to-end encryption, or embrace privacy-by-design (that is, the data on a device is private by default and only shared when approved by the user, rather than being open by default). Apple stands out here by promoting its ecosystem as enhancing privacy. Although the company is data-hungry – its intellectual property evolves by understanding how users engage with devices – selling ads is a minor business compared to the profits it makes by selling apps, content (for example, music and books), and devices. As such, it can afford to roll out and emphasize privacy-enhancing features such as its Intelligent Tracking Prevention (which offers users new levels of privacy while on the internet), although advertising firms are intent on finding workarounds.[19] Apple has also promoted encryption as a way to offer users more privacy. As open source platforms grow in popularity, proprietary systems might have to change to privacy-by-design models to retain paying customers.

Finally, a development to note here is the so-called 'decentralized web' (DWeb). For some industry players, such as world wide web pioneer Tim Berners-Lee, core features of the web, including pervasive data extraction and monopolistic proprietary platforms, demonstrate the flaws of a centralized web which has become dominated today by a small number of enormous and powerful corporations.[20] They argue the ground on which surveillance capitalism has emerged can be disturbed and turned over if a decentralized web takes shape. The crux of the matter is enabling users to create and move data on their own terms without necessarily relying on, or even interacting with, proprietary systems and architectures. The point of building the DWeb, then, is to create new protocols and platforms that can deliver control to the decentralized user and protect their privacy when they create and consume data. For example, instead of the current model where Google or Facebook own gigabytes of data about each user, centrally stored on their servers, the DWeb would hold data locally, sharing it through peer-to-peer connectivity.[21] Rather than using a system such as HTTP (HyperText Transfer Protocol) to find data at a fixed location, the DWeb would find information based on its content, meaning it could be stored in multiple places at once.[22] A user inside the DWeb would retain control over and possession of their data. This would have the effect of breaking down centralized databases, making it more difficult for third parties to access and mine these. Google or Facebook could engage with our data or briefly analyze parts of it for clues about how to deliver a service, but they would no longer own it for perpetuity, nor would they be able to freely and endlessly slice through it to find clues for advertisers or product development.[23]

Achieving this sort of DWeb vision, which we think is entirely conversant with the principles of slow computing, won't be easy. In the first place, if there is insufficient user adoption, there won't be enough developer opt-in or business viability, so any apps or programs will struggle to become mainstream.[24] Then there are questions about whether the big tech players will welcome or (as might be more likely) fight against DWeb applications, especially if they truly undermine the basis for data extraction. Finally, authoritarian governments (and even 'liberal' regimes accustomed to their new powers of surveillance) won't like the idea of a DWeb. As users of encrypted messaging apps such as Signal and Telegram have discovered, some governments are intent on cracking down on forms of communication that security services cannot monitor.[25] An encryption-rich and mainstream web operating beyond the state would likely fly in the face of government policies that favour data retention for the sake of surveillance and criminal investigation.

The role of government

Government has a dual role when it comes to data extraction. On the one hand, it is a significant generator of data (often excessively so), which it seeks to use to manage and govern society. In this role, it often procures private services to fulfil its mandate. On the other hand, it has the role of holding companies and others to account with respect to regulations and laws regarding the capture, processing, and use of data. In both cases, in representative democracies governments are meant to fulfil the will of the people. In other words, government should act to serve citizens. If we desire better privacy then our governments should implement new policies and laws, and direct the actions of the civil service to ensure that these are met. After all, we elect governments and we can also vote them out of office if they fail to deliver on expectations. Our protests, lobbying, and voting patterns work to promote certain policies. Of course, government also tries to foster economic development and prosperity. And, as we have discussed, business interests in the digital realm can clash with what society as a whole might need. Moreover, there might be groups in society who favour different approaches to issues. Politicians and civil servants must navigate a complex and often contradictory landscape when formulating policy and regulation.

Governments have obligations with respect to how they manage data, with some key laws written to protect privacy and unscrupulous use of data. In recent times, the Fair Information Practice Principles (FIPPs), first published by the OECD in 1980, have formed the cornerstone of privacy and data protection laws across a number of countries. FIPPs consist of eight principles setting out requirements regarding the generation, use, and disclosure of personal data and the obligations of data controllers:

Notice	Individuals are informed that data are being generated and the purpose to which the data will be put
Choice	Individuals have the choice to opt-in or opt-out as to whether and how their data will be used or disclosed

Consent	Data are only generated and disclosed with the consent of individuals
Security	Data are protected from loss, misuse, unauthorised access, disclosure, alteration and destruction
Integrity	Data are reliable, accurate, complete and current
Access	Individuals can access, check and verify data about themselves
Use	Data are only used for the purpose for which they are generated and individuals are informed of each change of purpose
Accountability	The data holder is accountable for ensuring the above principles and has mechanisms in place to assure compliance.[26]

Many of these principles have been eroded in the big data age. This is partly to do with the advances and pervasiveness of new technologies, partly because data controllers and processors have wilfully ignored them, and partly because of active lobbying by the data industry to limit their liabilities and responsibilities and extend the value they can extract from data. The fact that FIPPs are now difficult to apply in practice and are routinely being circumvented has highlighted the need to revisit and revise them. This need has been recognized in the EU through its new General Data Protection Regulations (GDPR), as well as in a number of other countries, such as the US, Canada, and New Zealand.[27] For example, the 'Consumer Privacy Bill of Rights' in the US set out revised FIPPs for the big data age, though the bill was never enacted.[28] FIPPs and attempts to legislate for them are a key way in which government can help enact slow computing with respect to data extraction.

There are numerous ways for government to act on our behalf and we can and should encourage them to do so. One way is by championing privacy-by-design. While regulatory and legislative compliance seeks to ensure that vendors and cities fulfil their obligations with respect to privacy by correctly and fairly handling the data they generate and manage, privacy-by-design proposes that privacy is the default mode of operation.[29] Rather than the default being the collection of data and the assumption all data are available for use, in privacy-by-design all data remain private unless the user explicitly says otherwise. In other words, privacy is hardwired into the design specifications, infrastructure, business practices, and usage of the technology.[30] The use of privacy-by-design has been advocated by the EU, the US Federal Trade Commission, and a number of national information/data protection commissioners.

More recently, privacy-by-design has been embraced and incorporated into the EU General Data Protection Regulations. Since May 2018, companies which conduct business within the EU have been required to think seriously about how digital devices or services produce data and how those data will be protected at all stages so that the user does not unwittingly create flows that companies can exploit. A core aspect here is the notion of privacy-by-default, which requires that companies conduct 'privacy impact assessments' to ensure that privacy does not somehow get in the way of how devices or services function.[31] If properly enforced, this is a significant move that bolsters the potential for practices akin to slow computing to develop.

This particular element of the GDPR also demands attention because it demonstrates what collective moves toward slow computing can achieve. Pressure for greater levels of data sovereignty moved from civil society through the European Data Protection Supervisor, the European Council, the European Parliament, and the European Commission, resulting in a seismic shift in data protection policy which, in theory, offers EU citizens a degree of immunity from the worst excesses of corporate data extraction behaviour. The big test will be whether data extraction companies are able to circumvent the GDPR. Only time will tell. If there is a suspicion of non-compliance, data protection authorities in each member state of the EU can investigate and punish companies.

Public services need to take active responsibility for the systems they procure for citizens, dictating more stringently what private companies running services on behalf of government can do with any data produced. For example, if a company develops an app for a public agency, the agency should also set out the terms concerning data generation, ownership, storage, security, sharing, conjoining with other data, and further use, and also what happens if there are data breaches. This might seem like common sense, but many governments are more interested in whether the service procured is delivered, not about any secondary capital which the company might derive from monetizing the data, so these conditions should not be overlooked. The desired terms should instead automatically form part of the procurement evaluation process. For systems that are already deployed, they should be assessed as to how closely they match the desired parameters, and a roadmap for achieving future compliance formulated as necessary. This is particularly important for city authorities, given that individual notice and consent are all but impossible for many smart city technologies.

Moreover, each public agency should have in place data protection governance and management mechanisms, such as data protection officers, privacy teams, and external privacy advisory boards. A privacy audit team would undertake privacy impact assessments of an agency's systems, liaise with departments about the privacy implications of their work, and coordinate staff training on data privacy and protection. A privacy advisory board would oversee and audit the work of the privacy team, advise on the work priorities and programme, ensure that response and mitigation plans and processes are in place, and that data practices and policies are clearly communicated to the public. For example, Seattle established a privacy advisory committee of external experts in the wake of criticism about data practices and public outcry over the secret tracking of citizens.[32] It is tasked with formulating and overseeing a set of ethical principles with respect to the use of digital technologies by city authorities.[33] In these ways, government shapes the privacy and data protection landscape through its contracting procedures and parameters, and its management and governance arrangements.

Finally, government can take an active role in promoting slow computing by providing education and training programmes. These need to be directed to different constituents. A key target group is the general public, with programmes designed to improve citizens' knowledge of what is occurring, inform them of their rights with respect to data extraction, and provide them with practical skills to protect themselves against privacy harms. This should be complemented by an educational programme aimed at schoolchildren, that warns them about data extraction across their digital devices and informs them about how best to manage their data privacy. Public sector staff would constitute the third targeted group, with programmes designed to inform them of their data protection obligations as a key data controller and processor. There should be an active training programme aimed at technology companies – and in particular start-ups and small-to-medium sized enterprises that do not have the in-house capacity for privacy expertise – to set out their obligations and best practices.

The role of community, civil society, and non-governmental bodies

While industry and government are key actors in both data extraction and its regulation, there is much that we can do collectively as civil society to enact slow computing. At the very least, we can hold government and industry to account and lobby them to protect our rights and treat us with respect. Beyond that, we can leverage the diverse skills within the general public – including the technical skills of coders, security specialists, and engineers, and the domain knowledge and skills of lawyers, attorneys, and lobbyists – along with the organizational capacity and political leverage of community organizations and non-governmental bodies (NGOs) to challenge dubious practices, create appropriate campaigns and social movements, reappropriate the technologies of corporate power and municipal governance, develop alternative technologies and interventions, and promote slow computing.

A collective move toward slow computing in everyday life might appear quite ambitious, and it does introduce some rather stark new complications, but it seems to be exactly the type of action needed. On the one hand, we can seek to enact an affirmative engagement with computing that is productive and set more on *our* terms.[34] This is about seeing the power and joy of computing as an opportunity to create social change. The problem is not digital technologies and data per se, but how they are employed: the power of computing should be reappropriated for the public good, rather than for corporate profit, state control, or individual gain. On the other hand, we can actively resist the more harmful and pernicious effects of acceleration and data extraction. The task here is not to position computing and a data-driven world as a threat to values, freedoms, and activities. It is to create oppositional practices and technologies that counter or provide alternatives to the services and platforms controlled by powerful vested interests. These two approaches map onto the proactive and reactive forms of data activism detailed by Stefania Milan and Lonneke van der Velden.[35] Taken together, proactive and reactive interventions seek to reimagine our digital society and recast it so that using computing is empowering and emancipatory, and its production and use is embedded within appropriate regulatory regimes and framed by democratic agency.[36] By working together in proactive or reactive ways, we can collectively lobby for and intentionally enact social change.

Quite like the example of building your own cloud, which we referenced in the previous chapter, there is scope today for collective moves to build community-run computing infrastructure and develop open source software which are based on a different ethos and are not bound up in data extraction. Framed as 'civic tech',[37] these kinds of initiatives take a do-it-yourself approach to creating the means for digital life. In essence, they take back the means of production and root them in community and not-for-profit ventures. And they are a perfect way to balance the joy of computing with slower values.

With respect to infrastructure, some of the best examples to consider involve efforts to create community-run Wi-Fi networks.[38] In some cases this reflects the failure of utility companies to extend connectivity. In Mankosi, in Eastern Cape province, South Africa, university researchers have worked with locals to establish a wireless community network under the auspices of a rural cooperative. The Zenzeleni project is a community-owned internet service provider (ISP). It delivers connectivity at lower prices than any mainstream ISP might offer, not least the mobile phone companies which charge exorbitant rates for data usage.[39]

In other cases, however, community 'mesh' networks emerge not only as a response to limited access or high costs (Detroit is one example[40]) but also as a way to protect communication, not least during protests or uprisings, as occurred in Egypt and Sudan during the 'Arab Spring'.[41] FireChat is an app that creates a community mesh between users that can be used to share information without intrusion from governments.[42] It is worth noting that in the background here are not only the technicians who create software such as the Commotion Wireless tool[43] but also institutions, such as the Open Technology Institute, that offer them support. Such collective efforts to create community-run digital infrastructure can successfully manipulate extant arrangements in ways that undercut data extraction by building in privacy from surveillance or government control or by ensuring that actors in civil society can continue to push the boundaries of what's possible in the digital age.

Perhaps one extension of this ethos will be platform co-ops that could operate in certain sectors of the economy, such as child care, and thereby facilitate everyday life to remain digital but without necessarily creating data for the tech sector.[44] A platform co-op is tantamount to a software mesh that can enliven and empower digital workers to control and develop aspects of their local economies. If platforms continue to be as central as they have become in the last few years, efforts to mesh talents and move resources via cooperative practices deserve to be seen as one element in the move toward a slow computing world.

One example of a platform co-op is the Grower Information Services Cooperative (GISC), founded and owned collectively by farmers in the US to provide an alternative to the data extraction practices of large agri-business firms.[45] GISC claims to be the agricultural sector's 'only grower-run data cooperative'.[46] GISC has now joined with the Agricultural Data Coalition, which was formed by farmers and allies in law, business, engineering, and academic research. Together they have established AgXchange which is 'designed as an open, central point of access for farm data'. It is, they claim, 'a neutral platform for the grower to control how their data is shared'; an initiative 'dedicated to creating the agriculture industry's first cloud-based platform that will be controlled by growers and open to all industry service partners and technology providers'.[47] In short, AgXchange is about finding ways of stopping data extraction and establishing cooperative data sovereignty.

Beyond community infrastructure, there is a strong case to be made for community use of encrypted internet platforms. For example, workers could push their employers to install Tor as the standard browser across company computers. Users can always pop back into the standard browsing environment if their bank's website or their favourite news site doesn't work as they might like, but normalizing the use of a Tor browser and including discussions of its merits in everyday conversation can help move slow computing from an individual to a more collective practice. If you are a member of a union, for example, raise the issue of privacy from data extraction as a core concern in the workplace and propose rolling out Tor browsers. If you manage or own a business, encourage your employees to obfuscate (in the ways suggested in Chapter 4). If you are an educator, use Tor when you browse online in front of students, for example when you demonstrate how a feature of your library's website works. Adopting platforms together with others, in a collective move, can make an effective difference, as well as feeling less isolating.

Civil society organizations are also important actors in the production of software, especially open source initiatives. These initiatives rely on the mostly volunteer labour of ordinary citizens to create and maintain the software, or add data to the platform as in the case of Wikipedia and OpenStreetMap. By working collectively, often at a distance from one another, digital products are created that have a different ethos to their commercial equivalents which are reliant on data extraction. While many of these projects are driven by a relatively small set of motivated and skilled individuals, others rely on crowdsourcing a substantial number of contributors. Linux has been built by thousands of contributors, Wikipedia consists of nearly six million articles written by thousands of writers and editors, and OpenStreetMap has been produced by mappers across the entire planet. There are plenty of opportunities to get involved in collectively producing open source forms of slow computing. Indeed, the larger the crowd, the more the tasks can be subdivided and undertaken efficiently. The willingness of many people to voluntarily participate in these projects is because they provide effective platforms to connect socially, communicate meaningfully, and contribute collectively to tackle common issues.[48]

Other kinds of civic hacking include hackathons and data dives that are usually more locally focused. Most hackathons (events which bring people together to design and build tech solutions) are seeking to address particular issues and develop new commercial products that have little to do with data extraction.[49] There are also more civic-minded hackathons linked to trying to address social issues (for example, Hack4Good[50]), or data dives that seek to use data for social good and enact data justice (for example, those organized by Data for Black Lives[51] and Detroit Digital Justice Coalition[52]). These more socially orientated initiatives express a slow computing ethos by building on the joy of computing to challenge the dominant profit-orientated business model of mainstream computing.

With respect to data extraction, a crucial type of software to consider is privacy enhancing technologies (PETs). PETs seek to provide individuals with tools to protect their personal data and dictate how they should be handled by different services.[53] PETs seek to minimize data generation, increase individual control of personal data, choose the degree of online anonymity and linkability of data, track the use of individuals' data, gain meaningful consent, and facilitate legal rights of data inspection, correction, and deletion.[54] PETs include relatively simple tools such as ad blockers, cookie blockers and removers, malware detection and interception, site blocking, encryption tools, and services to opt out of databases held by data brokers.[55] Many PETs are developed by civic society organizations, such as the Electronic Frontier Foundation.

A core dynamic in the battle with data extraction is going to involve actors in civil society trying to trace the flows of data through devices and services to pinpoint illegal actions. Data processing – grabbing, mining, and marketing – is opaque. A number of digital rights organizations seek to expose illegal data extraction practices and to produce tools to protect users. The London-based NGO Privacy International (PI) is one prominent player. PI dedicates part of its resources to exposing how digital society really works, for example by preparing illuminating reports and case studies on issues such as connected cars, smart cities, and free Wi-Fi, and using them to lobby industry bodies and governments. But it has also been active in shaping policy and legislation. With respect to the EU's GDPR, which we mentioned earlier, PI's first submission to the EU was in 2010. It continued to track and make submissions, established alliances with other NGOs in Europe, and tracked when MEPs were proposing amendments and comparing them to submissions or suggestions from industry. As PI have noted, they 'waded through thousands of amendments and compromise amendments, and drafted and suggested our own to persuade

Members of the European Parliament (MEPs) to adopt them; we participated in many meetings with MEPs and their political advisers, attended and spoke in committee hearings, developed blogs and educational resources'.[56]

PI also contributed to European Digital Rights (EDRi), an overarching network of around 40 NGOs from across Europe. Appropriately enough, its base is in Brussels. From there it conducts analysis and establishes links with MEPs to monitor EU policy development. EDRi was a crucial player in the GDPR process. Its conclusion, like many other digital rights advocates, is that the regulation is 'not perfect [but] is probably the best possible outcome in the current political context'.[57] Or, as Anna Fielder, Chair of PI put it:

> ... data-hungry companies and governments, and poor technology designs continue to make our personal data vulnerable [but now] we have a legal instrument to hold the powerful to account. We are going to use this legal regime to help empower citizens and consumers. And we are going to test it against emerging business models, ambitious and delusional government programmes, and any system that takes control away from the individual ....[58]

Similar battles around data extraction are occurring across Europe, where campaigners for digital rights and privacy are raising the alarm over emerging surveillance infrastructure. In the UK, for example, the Investigatory Powers Act 2016 (also known as the 'Snooper's Charter') would require ISPs to record one year of internet connection records – essentially a list of websites users have visited – and enable a vast array of government agencies to see those records. The civil and human rights organization Liberty (alongside Privacy International) has been a major critic of the Act. Here, the significance of contemporary collective action in support of practices akin to slow computing becomes particularly evident. In early 2017, for example, Liberty received just over £53,000 in crowdfunding to launch a legal challenge of the Act in the High Court. The challenge was successful: in April 2018, a judge ordered the government to amend the Act so its focus is only on serious crime.[59] Liberty has also challenged other aspects of the Act. This experience demonstrates that organizations like Liberty – with expertise and ability, matched with today's digital tools, such as crowdfunding to connect them with the wider public and raise needed funds – are essential components in creating a legal architecture that protects slow computing practices.

Some additional clues about the ongoing and emerging role of civil society in the digital age are provided by dwelling briefly on the work of the Electronic Frontier Foundation (EFF). Based in San Francisco, EFF advocates for greater levels of online privacy and frequently leads campaigns around pertinent legislative issues. For example, in 2018 the EFF 'action center' campaigned against a bill in the US Congress that would introduce biometric screenings within the US air network. Another campaign spotlights attempts by intellectual property holders to eliminate privacy when registering a domain name for a website. In the background of these campaigns is extensive and advanced knowledge about, and ongoing analysis of, how digital society operates. One core element in this work is EFF's team of IT security specialists and attorneys, who piece together an understanding of the links and flows that strip away our privacy and facilitate surveillance. In addition, EFF makes PETs to help people protect their data, such as Privacy Badger, a tracker-blocking tool.

A final example of collective civil society efforts against data extraction centres on the American Civil Liberties Union (ACLU). Based in New York, with offices across the US, the ACLU builds on the Bill of Rights of the US Constitution to defend and preserve the individual rights and liberties of US citizens. The development of digital technologies introduces particular challenges, especially regarding privacy and speech. In its 'Speech, Privacy, and Technology Project', the ACLU's lawyers and policy analysts work on diverse but crucial cases that address a wide range of technological developments. They question the legality of introducing electronic tolling on highways that 'would capture sensitive information about millions of drivers'.[60] They interrogate the extent to which facial recognition software is used in the retail sector.[61] And they draw on the ACLU's experience and use of instruments such as the Freedom of Information Act to discover how 'agencies like the FBI and the Department of Homeland Security are collecting and analyzing content from Facebook, Twitter, and other social media sites'.[62] In short, the ACLU works to lay the ground on which slow computing practices can emerge under the protection of the Constitution. This is the type of collective effort – 'collective' because the ACLU draws on the public for funding and political support – needed in every jurisdiction around the world.

A slow computing world

The discussion here and in the previous chapter has examined a variety of individual and collective moves that could facilitate slow computing. A more balanced digital life and society are within our grasp, although acquiring and holding on to these are not straightforward. As we have reiterated, we celebrate today's digital technologies but we do so with the caveat that certain features and practices are extremely problematic. Truly appreciating the joy of computing means holding on to some control over your time and participating in digital society without the levels of data extraction and surveillance we see currently. We can enter into the age of slow computing if we work hard enough to restrict the sorts of forces that compel tech companies and encourage governments to construct architectures that disempower digital subjects and undermine democracy. The benefit of a collective approach is that any moves towards slow computing are shared by all, not simply by those who have adopted slow computing practices or philosophy. By clubbing together, we create a critical mass that can produce change that no one individual could do on their own.

However, we cannot rely on others to do this collective work on our behalf. We need to get actively involved in whatever ways we can. That doesn't need to be in a leading role or as a gung-ho frontline advocate. It could be via lobbying politicians, taking part in civic hacking, or joining or supporting a civil society group or NGO. If we don't get involved, then we risk allowing digital life to unfold and its shape to be decided by others, principally the companies who stand to gain from datafication, surveillance capitalism, and the rolling out of the psychological tricks and manipulations we reviewed in Chapter 4.

Further, trying to undermine the normalization of data extraction as a whole can assist in making it harder for companies and governments to pursue these intrusions. One analogy is testing cosmetics on animals: once a standard practice within the cosmetics industry, it has become deeply unpopular with consumers, and this exerts pressure on companies to change their practices. Extraction is standard practice today, but chipping away at its legitimacy can create the conditions for whistleblowers to come forward in the future and shed light on where it is still taking place. An embrace of slow computing as a collective endeavour to delegitimize data extraction has a role to play in reducing our overall exposure. There is a potentially much broader societal impact in recognizing that we produce data, that data become commodities with a significant value, and that our computing lives in the future should not by compulsion contribute to the reserves of data established by companies and governments.

In our view, the future shape and nature of digital life is still very much in the making. There are powerful actors producing seductive new technologies that will continue to excite. The joy of computing is emergent. How we will connect with the technologies of the future and how they will shape our lives is at issue. Slow computing asks us to question how we live with technology. In this regard, therefore, it is necessary to consider some of the philosophical issues in play today and to be resolved (or not) tomorrow. As we explore in the next chapter, the dynamics of slow computing raise a slew of normative questions about what our digital society and economy *should* be like. Addressing these questions – and finding answers that work for us as individuals and as collectives – requires critical reflection and an astute analysis of what kind of society we want to live in and how we can ensure it is realized. For us, that means setting out the philosophical basis of slow computing that coherently articulates its logic and rationale, and it is to this that we now turn.

6

An Ethics of Digital Care

Saturday. 11am. You've got withdrawal symptoms. You haven't touched your smartphone all morning. The laptop is closed. The kids' games machine is turned off. You really want to check your emails, to see if anyone responded to your latest Facebook post, to catch up with what's happening in the world. You think about taking a quick peek, but your partner reminds you that it'll all be waiting for you later; that none of it is going to disappear; that your family is more important. Usually, you would ignore all this and reach for the phone in any case. Today, however, you turn it to silent and head off for a family trip to the park, where you have a great time untethered from the digital leash. You realize it actually feels good to forget about being online and constantly fretting about who thinks what, or what needs to be done. You'd still like a sneak peek, but you're determined to resist and enjoy a lazy lunch.

We need a new ethics of digital care. We need to adopt and enact a philosophy of slow computing – of stepping back and uncoupling the digital leash to reestablish control over our time and our digital footprints and shadows. We stand to experience the joys of computing, while minimizing some of the more pernicious aspects of the emerging digital society and economy if we pursue individual and collective, practical and political, actions. In this chapter, we extend our overall argument by developing the philosophical underpinnings of, and for, slow computing. The target is a rationale and imperative for adopting slow computing; for its practice not just by individuals but also by wider society, including companies.

While we could practise slow computing without embracing slow philosophy or becoming members of a slow computing movement – simply adopting some actions as they suit – our practices gain meaning and maximize their impact if they are rooted in a coherent set of ideas and ideals that justify and promote them.

At issue here is understanding the key normative arguments for slow computing. By 'normative' we mean a type of thinking concerned with how things *should be*, not how they presently are. It is about seeking answers to questions such as: What kind of digital society do we want? How should it operate? How should we go about creating it? Of course, how an issue is framed normatively also shapes how it is approached and any subsequent debate. For example, a position that accepts the need for mass surveillance and the erosion of privacy will advocate for different interventions to address any public concerns about data extraction than a position that is much more committed to individual rights and personal autonomy.

For us, the principles of slow computing are normatively rooted within an ethics of digital care – self-care and collective care – designed to reimagine and remake our digital society and economy so that it protects and enables personal and societal interests. There is an established philosophical literature on the ethics of care, rooted in the ideas and ideals of feminism, that promotes moral action at the personal and collective level to aid oneself and others.[1] As an ethical approach, the ethics of care is relational in nature – we care for each other in many different ways – recognizing that we are bound together in webs of relationships that bring with them responsibilities, obligations, and duties. Some of our care is reciprocal, in the sense that how we act towards others is reflective of how we expect them to act towards us, but much of it is non-reciprocal, taking on care because we are obliged (acting as a parent, friend, worker, or employer) or are acting altruistically (such as volunteering). An ethics of care therefore recognizes that forms of care, and what kinds of care are required by different people, vary by context and circumstance; there are no one-size-fits-all solutions.[2] What works best for you and your family and community might not be optimal for ours.

This approach is in contrast to other ethical theories, for example relating to social justice, that seek generalizable standards, rules, and principles, often expressed in terms of rights, entitlements, and responsibilities.[3] This difference is reflected in how a matter is framed in terms of a question: an ethics of care approach tends to ask 'how best to respond?' in order to address an issue, whereas other ethical theories tend to ask 'what is just?'[4] Put another way, the first question demands action, whereas the latter is an instruction not to treat people unfairly. Joan Tronto, a care ethicist, suggests that there are four key elements of care: attentiveness – recognizing the needs of oneself and others in order to be able to respond; responsibility – taking it upon ourselves to take action; competence – having the knowledge and skill to deliver on one's responsibilities; responsiveness – that those being cared for are able to adequately receive the care.[5] Slow computing is about making sure all four elements are in place with respect to digital life: that we care for ourselves *and*, just as importantly, for others.

With respect to the latter, non-reciprocal care for others might involve helping those who lack digital literacy, by taking on some of the digital burden of setting up and maintaining digital technologies to enable them to live digital lives. It could involve seeking environmental justice for those affected by the mining of metals needed to produce digital devices or dealing with digital waste.[6] It might also mean caring for our wider environment in terms of enacting sustainable computing with respect to energy consumption (which contributes to climate change) and resource extraction (which endangers local environments and ecosystems). Digital technologies are significant users of energy and resources, and slow computing is a step (albeit a modest one) towards the decarbonization necessary for creating a more viable future.[7]

We note that an ethics of digital care could seem counterintuitive from some political vantage points. For example, if you are a free-market capitalist, why would you care about worker wellbeing as long as you can extract profitable value from them? However, regardless of whether your political leanings are to the Left or the Right, whether you're an ardent capitalist or passionate socialist or any other political hue, you can hopefully recognize the value of an ethical commitment to looking after your own and your family's wellbeing, as well as that of your community, customers or employees, and the wider environment. The reason why well-known German companies have implemented slow computing practices, such as the right to disconnect (see Chapter 5), is that they have realized having less-stressed workers improves productivity, promotes innovation, reduces employee turnover and lost days, and increases profit.[8] Other firms have realized an ethical approach toward customers creates brand loyalty, hence why they promote their environmental sustainability or corporate social responsibility practices and initiatives. Caring can be altruistic and good business sense, at the same time. However, it is important to call out what has been termed 'green washing'[9] and 'ethics washing';[10] that is, companies claiming that they are acting in a caring, sustainable, and ethically way, when their actual practices reveal this not to be the case. As noted in the previous chapter, we are also not divorcing an ethics of care from more universal claims for rights and entitlements that protect and provide care at a general level. Legislation related to working hours and privacy that work universally can play an important role in creating opportunities to practise slow computing and enabling care to take place.

Our vision for a slow computing ethics of digital care, which we go on to outline in this chapter, consists of two primary pillars. The first is about an ethics of time sovereignty and we propose two key aspects: a focus on the present and an ethics of deceleration, disconnection, and asynchronicity; and a focus on what lies ahead and an ethics of the future. The second pillar is about an ethics of data sovereignty, which also has two key aspects: a focus on extraction and an ethics of privacy and forgetting; and a focus on how data are processed and value created and an ethics of artificial intelligence and data usage. Having outlined and discussed these pillars we then consider whether a slow computing ethics of digital care can extend to all, regardless of gender, race, class, age, and literacy, or whether there are some who will be able to more easily practise and gain from slow computing than others.

Ethics of time sovereignty

In Chapter 4, we defined time sovereignty as the power and autonomy to dictate how our time is spent. We recognized we can never have full sovereignty over our time because we are bound into social and institutional obligations and responsibilities – family time, working hours, club and society schedules, and so on. Nonetheless, within these constraints, time sovereignty gives back as much autonomy as possible. Here, we want to extend this idea by considering it in relation to an ethics of digital care, not just for the here and now but also for future generations.

Ethics of deceleration, disconnection, and asynchronicity

As we detailed in Chapters 4 and 5, central tactics for practising slow computing are deceleration and disconnection. An ethics of digital care in this respect is a commitment to enabling those people who want to embrace slow computing to downshift the pace and tempo of everyday life; to reject timeshifting and the need to fill 'dead time' with seemingly more productive activities; to cull the pile-up and interleaving of tasks that creates temporal densification and fragmentation; and to forego always being open to on-the-fly encounters. Such an ethics seeks to allow people to take greater control of their use of time; to feel less harried and stressed; and to achieve a work-life balance. To allow people to move back to natural, social, and clock times, rather than living always within the frame of network time. And to not feel guilty or be penalized for doing so.

Beyond facilitating self-care, a temporal ethics of digital care also extends to a broader commitment of societal care – of practising care for others through changing working conditions and workplace policy, and protecting people from harm through regulations and legislation. Care here is almost akin to public health; it is changing social conditions for the benefit of all. There is an acknowledged and acted upon moral imperative to tackle issues such as: working time drift; the expectation that workers are always-on/everywhere-available and are receptive to immediate response; the erosion of vacation breaks and periods of rest; the normalization of overwork, stress, and burnout; and the framing of being a workaholic as a badge of honour. The imperative is to value: quality over quantity; the enjoyment of undertaking a task over the speed with which it is completed; and performing a single task well rather than multiple tasks poorly. It is about the imperative to recognize the value of rest and wellbeing over tiredness and stress, and that these have societal and business benefits, not just individual ones.

Another temporal aspect that requires an ethics of digital care is the adoption of real-time systems. While operating in real-time potentially provides efficiency and optimization, it also provides no time for reflection, contemplation, and deliberation, and shifts decision making from people to algorithms. It trammels citizens and managers alike into narrow options and automation. In turn, certain forms of knowledge, values, and ways of doing things become prioritized over others. For example, *techne* (instrumental knowledge which is more easily encoded) is prioritized over *phronesis* (knowledge derived from practice and deliberation) and *metis* (knowledge based on experience).[11] Such a transition has consequences to how everyday life is experienced and governed by making it more technocratic, less open to debate and deliberation, and downplaying or eliminating the relevance and lessons from the past and the lived experience of whatever is being managed in real-time. As others have started to argue, there is merit in a counterview of valuing asynchronicity and temporal dissonance, of ensuring that we can live at our own pace, not just slowing down but operating at differential speeds.[12] Here, an ethics of asynchronicity works to prioritize people, their activities, and their wellbeing and desires, as well as factors such as neurodiversity, not simply the efficiency and optimization of systems.[13] Instead, we might imagine systems being designed so they have layers or routines that do not work instantaneously, are out of alignment and incongruous or decentralized, and are more speculative, poetic, and unexpected.[14] Real-time systems configured in such a way might produce more lively places, not ones that are ordered, rote, and passive.

The imperatives we have discussed here are ethical commitments because they are underpinned by judgements concerning what is right or wrong, fair or unfair, or just or unjust. There are issues of morality at play. If we accept there are harms arising from acceleration and yet we do not facilitate slow computing that works to ameliorate these harms then we are deliberately working against the interests and care of ourselves and others. Perhaps we can try and justify this position by arguing that any harms are compensated for in other ways, such as a salary cheque and bonuses for long hours and constant availability. But often, as with working time drift, this just leads to more exploitation, a squeeze on labour, and more stress and insecurity. Instead, ideas of fairness, equity, justice, citizenship, and democracy – key concepts in moral philosophy – can be mobilized to argue for and justify claims to time sovereignty. These are the conceptual tools by which labour unions have pursued improving working conditions, such as reducing the number of days worked and the hours worked each day, and gaining public holidays, vacation entitlements, and overtime payments. As the power of unions has dissipated in recent decades, campaigns to ensure time sovereignty have abated, just when digitally powered acceleration has been transforming working life. Perhaps now, in the age of network time, such campaigns need to be revived as in France, Italy, and Germany, with the ideas and ideals of slow computing at their core.

Ethics for the future

The ethics of deceleration and disconnection tend to be concerned with addressing issues in the here and now. They seek to change our present wellbeing by enabling us to exercise a degree of control over how we spend our time. But what of the future, both our own and that of future generations? Political and design decisions taken today can have very long-term effects that will be experienced not just by us but also by an open-ended chain of generations who currently have no voice, advocates to speak for them, or votes.[15] Surely we have a duty of care to, on the one hand, imagine and enact different ways of responsibly creating our future digital society and economy, and on the other to formulate slow computing so that it benefits not just us but also those that follow us. Adopting an ethics for the future recognizes that the relationship between digital technologies and society we make today produces a legacy that will impact posterity.

To help us map out an ethics of our relationship with the future, we draw on the work of future theorists Barbara Adams and Christopher Groves, and their book *Future Matters*. They detail two ways in which we approach the future, which they term 'future present' and 'present future', neither of which consider the future as a blank canvas yet to be created or already fated. Rather, the future is partially charted from the present (either predictable or forecastable based on past performance[16] or through plans and policies designed to produce particular outcomes) and, conversely, anticipated futures shape thought and action in the present (for example, concerns over the effects of climate change in the future are leading to policy changes in the present). The future remains uncertain, but we have a sense of what probable and possible futures might be realized, and also what preferred future we might aspire to.

The 'present future' is the future projected forward from the standpoint of the present and of past trends. It extrapolates forward what the future will or could be like from where we are now. The present future is not predetermined: while there is forward momentum, shaped by embedded structures, established path dependencies, and new innovations, the future is still uncertain and open to contestation and choice. It is then possible to imagine, plan, forecast, and create the future; to reshape how our own and others' future will unfold. We do this all the time. As in the notion that we are *homo prospectus*[17] (see Chapter 2), we are naturally future-orientated, concerned with what will and might happen over our life course, and often happiest and healthiest when we think about what lies ahead. Care implicitly concerns the present future – through caring in the context of existing situations and what is occurring, we seek to manage our own and others' path forward and create a better world for us to live in. Caring for one another, we attempt to judge what others want to achieve or need to thrive, and we try and help fulfil those ambitions.[18] We formulate strategies and plans, which we then try to enact to realize preferred futures. We do so, however, principally in, and from, the present. Practising slow computing is an ethically grounded way of shaping one's future.

Furthermore, practising an ethics of the future means countering the futures envisaged by others that reproduce and extend an accelerated world with extensive data extraction. Corporations and the state both evoke future imaginaries designed to preempt, prepare the ground for, and promote particular digital futures.[19] For example, smart city advocates produce slick videos and glossy adverts forecasting the potential benefits of a city saturated with ubiquitous computing. They actively lobby city and national governments seeking investment in new smart city schemes. They aim to enact what the geographer Ayona Datta terms 'fast urbanism': various forms of accelerated urban development, from fast-track planning to the adoption of real-time management systems.[20] Slow computing seeks to produce a more humanizing future smart city by trying to shift the thinking and practices of the present future towards slow urbanism.[21] This means turning the driving logic of creating and promoting smart city technologies away from speed, efficiency, optimization, and technocratic governance towards fairness, citizenship, social justice, and the public good.[22]

Caring in the present future is primarily about creating a difference that will be realized before the limits of our own mortality, not necessarily for future generations beyond our lives (though they may well inherit those outcomes). The 'future present' has a longer term vision that is more normative in orientation, seeking to envisage some state that we might wish to achieve in the distant future – for example, what a slow computing world might look like in 2050. It then uses the technique of 'backcasting' to work backwards to the present to try and define the steps or pathway needed to make such a future a reality. In other words, rather than working from the present forward, it works from a preferred future back. The future present thus acknowledges explicitly that our present actions potentially impact on future generations and that we need to act morally and ethically to create a different world.[23] A normative future is thus evoked to preempt, prepare for, or prevent threats from being realized, and to redirect present future paths onto a new trajectory.[24]

To revisit our example of the smart city, a slow computing approach to the future present would be to envisage what we think an ideal slow city should be like at some point in the future and then try to work out the steps needed over the intervening years to make that city. We're not concerned with the forward momentum of the present future; in fact, we are open to imagine any new means of reconfiguring technology we wish. The aim, after all, is to disrupt our present path and shift it in a new direction. By doing this, we free ourselves from focusing on surviving in the here and now and generating prefigured futures; instead, we can plot what we should do. In turn, we start to take responsibility for the future.

Indeed, a key aspect of an ethical approach to both the present future and the future present is taking responsibility for our actions and their consequences for future generations. This might seem like common sense, but while we might feel ourselves well intentioned, nearly all of us live lives that have potentially serious consequences for the future: as a planet we consume far too many resources and we try to ignore or defer issues such as climate change, species extinction, and excessive debt. Instead of taking personal responsibility, we cede responsibility for the future to others – business, government, civil society. In turn, business trades turnover and profit against the future in the narrow interest of the present. Similarly, government trades its short-term interests and getting reelected against the future. In effect, irresponsibility is institutionalized, which in turn encourages us to fly blindly into the future.[25] As a result, we often collude in policies and practices that will create sub-optimal, or even dystopian, techno-futures. Slow computing is about actively taking responsibility for the future, rather than ceding such responsibility. The forms of individual and collective actions detailed in the previous two chapters are designed to anticipate, prepare for, disrupt, and transform our digital lives and produce alternative futures.

But how can we act responsibly, with respect to unanticipated harms? Slow computing is predominately concerned with creating a future that addresses known and anticipated issues – of the world becoming ever faster, of privacy being further eroded, of data being used to extract ever more value. But what of unforeseen and unintended consequences arising from new technological developments and business practices, some of which might lie dormant for years until particular conditions exist? A slow computing ethics of digital care has to be reactive to emerging harms; this can be a problem given the speed at which new innovations are being invented and shipped to market. Responses always seems to be catching up and social, political, and legal interventions necessarily move slower given they need to be formulated, debated, and agreed upon, then implemented.

The solution to protecting the future forwarded by Christopher Groves is the use of a qualified precautionary approach to the development and adoption of new technologies.[26] At present, precautionary approaches generally use risk analysis to assess potential future harms; they view any concerns identified as issues to be tackled by further innovations that will ameliorate their effects, rather than to be avoided altogether. In other words, they act as a 'downstream brake', rather than taking a different route altogether. Groves suggests starting from a different precautionary position, critiquing the purposes new technologies intend to serve and assessing whether the innovation is likely to produce the public good and society we want future generations to inherit. One aspect of practising slow computing, then, is about embracing a precautionary analysis and acting on its results, aimed at creating a digital society posterity will thank, rather than curse, us for. This kind of work is already underway by civil society groups such as EFF and ACLU, as discussed in the previous chapter. And as we now discuss, this ethics for the future and this associated precautionary approach apply not only to time sovereignty but also to data sovereignty, protecting citizens in the future from excessive data extraction and harms associated with such activities.

Ethics of data sovereignty

Data sovereignty concerns authority and control over the generation of data about us, the data captured, and how those data are used. As with time sovereignty, given the state's need to use data to govern and the need for companies to keep records and use data to extract value, data sovereignty is contested and negotiated. In general, ordinary citizens have limited autonomy and power in terms of setting the terms of data sovereignty. However, as we have detailed in Chapters 4 and 5, we can adopt a range of tactics to try and shape what data are captured and campaign to introduce and change regulations and legislation that limits data usage. An ethics of data sovereignty concerns the moral arguments used to shape the terrain of data sovereignty. This concerns an ethics of digital care as well as more universal claims to justice, rights, and entitlements. An important emerging concept with regards to the latter is 'data justice'.[27]

Data justice draws much of its moral argument from the ideas of social justice. Just as social justice concerns the expected and acceptable ways in which people are treated, data justice concerns the fair treatment of people with respect to their data.[28] Likewise, just as social justice is underpinned by moral rights that set out what people can expect as members of a society (for example, in most of the Global North: freedom of expression, voting in elections, full recourse to the law, access to education and medical treatment, and so on), data justice is underpinned by a claim to moral rights such as those set out in the FIPPs (for example, notice, consent, choice, security, integrity, access, minimization, accountability) (see Chapter 5). It is important to remember that rights are not given and inalienable. Gaining them involves hard battles. Moreover, there are many theories of justice that set out different moral positions. For example, theories of social justice fall into four broad types: distributional (fair share); procedural (fair treatment); retributive (fair punishment for wrongs); and restorative (righting of wrongs).[29] Generally, data justice to date has been procedural in orientation. Here, we elaborate two sets of ethical arguments concerning data sovereignty and data justice, first in relation to data extraction, then in relation to how those data are processed and used.

Ethics of privacy and forgetting

At present, the debate over data sovereignty hinges on two trade-offs: between privacy and economic growth, and between privacy and national security.[30] In the first case, data extraction about customers is cast as a choice between creating new products, markets, jobs, and wealth *or* protecting individual and collective data rights. On the one side, there is the argument that privacy should not hinder innovation and the leveraging of economic value of individual data, or impede customer experience.[31] Without new innovations, some commentators and businesspeople suggest, the economy will stagnate and society will suffer. On the other side, there is the argument that it is possible to extract value from data and create new products without infringing on privacy and aggressively profiling and targeting individuals.[32] In the second case, data extraction is cast as a choice between creating safer societies *or* defending personal autonomy. On the one side, trust is traded for control, with citizens treated as potential threats to the wellbeing of a nation for the greater good of national security. In opposition, privacy and anonymity are seen as indispensable features of liberal democracies. Excessive data extraction erodes core societal values of freedom and liberty and enables the operation of more authoritarian regimes.[33]

Within both trade-offs, for those who are in favour of mass data extraction privacy is positioned as mutually exclusive from economic development and national security. Moreover, privacy is cast as dead or dying.[34] Even if there are those who do not fully realize it yet, it has been sacrificed for a greater good. This is accompanied by arguments such as, 'if you have nothing to hide, you have nothing to fear', or 'if you do not like how we operate, do not use our service'. As critics note, the first assertion conflates privacy with the concealment of suspicious behaviour, as opposed to personal autonomy, freedom of expression, and the selective choice to reveal oneself.[35] The second is entirely impractical and unreasonable given that digital devices and media are the tools of modern life, necessary for a career and social life.[36] Opting out is not always a viable choice. Indeed, with respect to many digital technologies, opting out is not even an option.

As we discussed in Chapter 3, we are all used to making these kinds of trades with our data, sacrificing privacy for a service. At times it can seem that privacy really is dead. And although the privacy debate is often framed in black and white terms, in reality it is full of greys: privacy is not yet extinct, even if it is under attack. Moreover, privacy is not mutually exclusive to economic development and national security but instead remains a significant value for many individuals, even if in practice they find it difficult to manage and protect.[37] It is still protected through legislative and regulatory instruments, even if these do not work as well as they might. The question here, then, is one of balance: of enabling a strong degree of individual data sovereignty while simultaneously enabling government to work and business to turn a profit.

For those seeking this balance, and asserting an ethics of digital care in relation to data extraction, privacy-by-design is important because it builds privacy into the very design of technologies, rather than layering it on afterwards through regulatory tools. The principle is straightforward: it argues that all data should be locked as private until an individual opens that lock. That opening-up does not consist of a blanket terms and conditions notice when an app is installed, it involves configuring the settings to unlock privacy settings. In other words, the app user controls their data sovereignty rather than blindly giving it away. This has been accompanied by civil society campaigns and legislative moves aimed at enshrining moral rights with respect to data sovereignty, as discussed in Chapter 5.

Back in 2007, one of us suggested a different approach to mass data extraction and long-term retention of data – an ethics of

AN ETHICS OF DIGITAL CARE

forgetting.[38] Our aim was to play devil's advocate with the drive to create technologies that produce, store, and manage a lifetime's worth of data about everything by suggesting that 'memory' (as data storage is often called) should be complemented by 'forgetting'. Daniel Schacter, a psychologist, identifies six forms of forgetting, three concerned with loss and three with error.[39] Loss-based forgetting consists of transience (the loss of memory over time), absentmindedness (the loss of memory due to distractedness), and blocking (the temporary inability to remember). Error-based forgetting consists of misattribution (assigning a memory to the wrong source), suggestibility (memories that are implanted either by accident or surreptitiously), and bias (the unknowing or unconscious editing or rewriting of experiences). We are mostly advocating for the use of transience and loss/erasure, but the other forms could also be woven into databases to provide a degree of imperfection, loss, and error.

Rather than seeing forgetting as a weakness or a fallibility, we framed it as an emancipatory process setting us free from our data footprints and shadows. We argued that there is a moral imperative to enable forgetting. This is to an extent built into data retention legislation that frames how long data should be retained, though data are routinely kept for very long periods of time, as we saw in our discussion of Google and Facebook's data holdings about us. While building erasure into digital systems seemingly undermines their integrity, it is a key way to ensure that people can put past mistakes behind them, evolve their social identities, live with their conscience, forgive others, achieve a progressive politics based upon debate and negotiation, and can ensure that an authoritarian or totalitarian state does not occur. As such, an ethics of forgetting works at both individual (being able to live with yourself) and collective (being able to live in a society) scales. As we will discuss in the final section of the chapter, these issues are particularly pertinent for marginalized and poor communities whose lives can be disproportionately affected by data-driven systems.

147

Ethics of artificial intelligence and data usage

Just as there are moral imperatives to practise an ethics of digital care with respect to the extraction of data, there are also imperatives with respect to how value is extracted from those data. Data are processed and analyzed using algorithmic systems. While some algorithmic systems run conventional forms of data processing and analytics which could be undertaken by hand, they are now using forms of machine learning in which the program learns from the data to refine its operations and analysis. Machine learning creates a form of artificial intelligence, in that the code is learning from its operation and can use this learning to improve its understanding and decision making. An important concern is the extent to which there are veracity and validity issues related to how these systems work. To what extent can we trust the accuracy, reliability, and robustness of the analysis and interpretation? Should you be concerned about how these systems handle your data and make decisions? Are they providing sufficient ethics of care? We think that there are some serious ethical questions that need to be asked and answered. There are two key issues of concern here: the data and how they are trained, and then the code underpinning these activities.

Any system is only going to be as robust as the data used to train the machine learning. If the data have issues relating to errors and biases then this will have effects on the results and interpretation. As a familiar refrain goes: garbage in will lead to garbage out. This is a significant issue. It is well known that all large-scale datasets contain issues of veracity (accuracy and fidelity), uncertainty, and reliability created through the measurement method and calibration, sampling frames, and data handling that introduce errors and biases.[40] For example, measurement is a process of abstraction and generalization and involves converting phenomena into some kind of representational form – numbers, characters, symbols, images, sounds, electromagnetic waves, bits – in which information can be lost or translated. How a method of data capture is set up, as well as the instruments used, can make a significant difference to what is recorded.

Moreover, data might be gamed or faked through the entry of false data (people deliberately giving untrue information) or the use of false accounts and bots. Data can be biased for a number of reasons. In relation to populations, the demographic sampled might not be universal, or could be skewed by gender, race, income, location, and other social and economic factors (for example, not everybody uses Twitter or Facebook, or shops in a particular store, or is on a particular phone network).[41] With respect to fidelity, it is not always the case that any set of data truly represents that which it purports to. This might be because a proxy is being used (for example, an R&D budget to represent innovation) or because of obfuscation (for example, people embracing slow computing and carefully curating their profile on social media rather than really presenting their true opinions).

These issues raise concerns over the trustworthiness, representativeness, and veracity of data, and raise questions about the usefulness of any analyses based upon them; such questions are not easily addressed by cleaning or wrangling the data. This in turn raises concerns about the profound effects on populations when they are represented by untenable data practices. The algorithms that process and handle data can further compound these effects. Algorithms consist of a set of defined steps that, if followed in the correct order, will computationally process input (instructions and/ or data) to produce a desired outcome.[42] Generally algorithms work using Boolean logic ('if this, then that') and the accompanying rule set, mathematical formulae and equations of calculus, graph theory, and probability theory. Producing a program consists of translating tasks and problems into algorithms and nesting them together to form a system. This process of translation can be challenging, requiring the precise definition of what a task/problem is (logic), then breaking that down into a set of instructions, factoring in any contingencies such as how the algorithm should perform under different conditions (control).[43] The consequences of mistranslating the problem and/or solution mean the program does not work as intended and can produce significant errors.

While processes of translation are often portrayed as technical, benign, and commonsensical, in reality their production is inherently framed and shaped by all kinds of values, decisions, and politics.[44] Programmers might seek to maintain a high degree of mechanical objectivity – being detached and impartial in how they work – but they can never fully escape the context in which the code has been created, such as local customs, culture, knowledge, and laws.[45] Programs evolve through trial and error, play, collaboration, reuse, testing, discussion, and negotiation: they are edited, revised, deleted and restarted, shared with others, and pass through multiple iterations.[46] As Tarlton Gillespie, a researcher with Microsoft, notes, a great deal of expertise, judgement, choice, and constraints are exercised in producing algorithms.[47] Moreover, algorithms are created for purposes that are often far from neutral: to create value and capital; to nudge behaviour and structure preferences in a certain way; and to identify, sort, and classify people.

In other words, digital systems are potentially full of all kinds of values and biases that can have a significant effect on how they operate and their outcomes. Sometimes these biases might be an implicit part of the system, other times they are intentional (for example, designed to process certain kinds of people in particular ways). Sometimes the system has the desired effect, other times there might be side-effects and unintended consequences. In proprietary systems (as opposed to those that are open source), which are generally the ones that have the most consequences, both the code and the organization producing it are usually 'black-boxed': that is, the code and how it was produced are not open to scrutiny and are therefore beyond query and question. There is clearly a data justice issue here in terms of being able identify, contest, and prevent any discrimination within these systems.[48]

An ethics of digital care with respect to both data and code is a moral imperative if we are to be able to probe, challenge, and correct any issues, to ensure that there is some level of oversight and accountability with respect to data-driven systems in relation to both their data and their functioning. This concerns more than privacy, data protection, and data integrity, which most regulatory approaches have focused on to date. As Nick Bostrom and Eliezer Yudkowsky detail, artificial intelligence systems need to be responsible in their actions, transparent in their workings, auditable in terms of their operations, incorruptible to manipulation, predictable as to their outcomes, and not make their innocent users 'scream with helpless frustration'.[49] Furthermore, there needs to be accountability in relation to the operations and outcomes of systems, with those failing to provide an ethics of care being open to redress and penalties for transgressions. Ensuring accountability requires a collective and institutional response given that most people do not have the literacy, skillset, or power to assess the integrity of the systems that affect them.

There is an urgent need, in our view, for a public debate about the ethics of machine learning, artificial intelligence, and data usage, and for the development of an effective set of responses that hold those creating and using such systems to account.[50] This needs to be accompanied by a wider normative debate, not simply about how these systems operate, but whether we want them at all, and if so, in what guise. If we are to create a slow computing world perhaps we need to set the parameters in which such innovation occurs, rather than simply reacting to technical innovations and their ill effects? So far as we see it, you should play an active part in shaping this debate.

A digital ethics of care for all?

In his book *In Praise of Slowness*, Carl Honoré asks: 'To what extent is slowing down a luxury for the affluent?'[51] We think this is a fair question. People who are wealthy generally have more control and autonomy over their schedules and lifestyle, and can offload work onto others. Those who are well educated are more likely to possess the literacy and skills to be able to configure their computers to combat data extraction. Consider us. Neither of us are rich, but we are both well-educated and in secure, permanent, reasonably well-paid public sector jobs with a relative degree of autonomy over our schedules. Neither of us have young kids consistently living at home, though one of us is a parent. We are both somewhat digitally literate, able to tinker around and perform the tactics set out in Chapter 4 with little help, though we're a long way from being able to make a living providing computer support. We are reasonably well disposed to be able to practise slow computing, to make a claim for time and data sovereignty.

But is that the case for everyone? Is slow computing equally achievable for other social groups? Are people of colour, women, immigrants, parents with kids at home, insecure workers, and those with low digital literacy easily able to practise slow computing? Conversely, are they more affected by the conditions of acceleration and data extraction? It seems to us that certain groups might be in a double bind: more caught up in the forces of speed and dataveillance, and less able to disentangle themselves.

Those with care duties have less autonomy over their time and are more likely to be tied to a digital leash. Their schedules are built around the needs and demands of others. They are more likely to be always on call and trying to deal with an ever-shifting diary as events are moved on the fly. They are running around after kids and their multiple activities – scooting from school to football practice to ballet lessons to play dates. They are checking in on older relatives, neighbours, friends – organizing home help, medical care, grocery shopping. They are trying to timeshift other activities between these – work, cleaning, cooking, shopping, and their own social activities. This disproportionally affects women, especially single mothers and mothers in dual-income households, more than men,[52] but all of us will be familiar with these scenarios. Slowing down and practising self-care while maintaining the same level of care for others is a conundrum. What's often required is a sharing of the burden and adopting scheduling tactics such as reverting to clock rather than network time. At issue is negotiating expected responsibilities and obligations.

Likewise, those who have insecure jobs – zero hours contracts, short-term contracts, non-tenured positions – are often expected to work as required, or feel they need to make a good impression in order to secure more favourable working conditions. They feel less able to ignore a call from their boss at 11pm or to turn down extended shifts and overtime. They are more likely to be performing piecework, where their rate of pay is directly tied to the speed they progress (for example, getting paid per kilo picked or number of items collected per hour). They are also more likely to be working in highly time-controlled and sanctioned environments. For example, some warehouses and factories use clocking-in/out systems and hourly targets, with penalties (including being fired) for lateness or missing goals, which in turn can then affect receipt of social welfare and finding other employment. The 'gig economy' – freelance, temporary work that can either provide supplementary income or become a main income, such as being an Uber driver, or hosting guests in your home, or doing piecemeal contractor work – pushes onto workers considerable responsibility for labour and the means of production (for example, the cost of a car or bike or computer), and alters the extent to which anyone can expect a salary.[53] In terms of supplementing an income, gig work also colonizes supposedly 'dead time', such as evenings and weekends, with work and its associated stresses. With bills to pay and income to earn, slowing down might not seem like a viable option. Similarly, self-employed people or the owners of small businesses need to hustle for the next contract and to deliver on time and within budget to turn a profit. This drive might be intensified if they are employing other people and the company's financial position is tenuous enough that slowing down might lead to laying off staff or the company folding (and thus failing to actualize an ethics of care for employees). Speed and responsiveness are bound into these working arrangements and slowing down, or trying to work 9 to 5, has consequences. Taking a holiday means receiving no income. Practising a digital ethics of care in such circumstances is not straightforward, and certainly more difficult to achieve than somebody working in a salaried, permanent position with defined hours, vacations, and conditions, perhaps supported by an active union, and with a pension waiting for them upon retirement.

With respect to data extraction and the uses data are put to, the evidence suggests that some people are more likely to have data captured about them and are more likely to experience the negative effects of profiling and social sorting. As we noted in the previous section, algorithmic systems are not neutral but have values inherently built into them and are designed to perform certain tasks. Decisions have been made about which data are captured, relating to whom, how these data are processed, what insights are garnered, what value is extracted and how, and what the consequences will be. There are choices about whose interests are represented within the dataset and whose interests are excluded. For example, state administration systems relating to social welfare, crime and law, income and taxation, health, and education inherently have value structures embedded within them that shape what data are captured.[54] These data are then used to assess entitlements and benefits and to monitor, govern, and sanction those citizens. Poorer and more marginalized people are much more closely tracked by the state: how they lead their lives is governed in ways wealthier people do not experience.[55] Likewise, they are prone to suffer when data brokers and retail companies deploy profiling systems.

For example, they are much less likely to be tagged as preferential customers who might be targeted with special offers or better prices or be extended credit; indeed, they are more likely to be redlined and denied custom altogether.

The question of data justice is about forms of data extraction that target poorer and marginalized communities via value structures within datasets and systems that propagate injustices and reinforce dominant interests. This is amplified with respect to people of colour and ethnic minorities, where algorithmic systems can exercise racial profiling, shaped both through the biases in the value structures within the code, and biases within the data that machine learning systems learn from.[56] Predictive policing systems in the United States for example have been critiqued for practising racial profiling and perpetuating institutional racism.[57] Both historically and presently black people are far more likely to be stopped and searched, to be arrested, and to be incarcerated. Both the data training and the ongoing data therefore suggest that black people should be further targeted. This then recreates a self-fulfilling cycle. The same accusation of racial profiling has been levelled at other predictive profiling and social sorting systems, for example related to credit scoring, customer relations management, and social services.[58]

In her book, *Algorithms of Oppression*,[59] Safiya Umoja Noble documents how racism and sexism are encoded into the data and algorithms of search engines and paid advertising. She reveals how search engines exhibit negative biases against people of colour, with similar search terms for black and white women and men revealing radically different results, with the searches related to black people more likely to be derogatory. For example, in relation to gender, a Google search for 'black girls' is more likely to produce sexually explicit results. When Microsoft launched its AI chatbot 'Tay' on Twitter, designed to get smarter as other users conducted conversations with it, within 24 hours it was tweeting racist and sexist remarks. Because its training data – those conversing with it – expressed misogynistic and racist views, the bot itself learned to express those views.[60] Given that this is how machine learning works, any domain that already has embedded racism and sexism within it will have it replicated in AI systems unless there are meaningful and concerted interventions.[61]

The implicit biases underpinning certain digital technologies can create far-reaching consequences. Consider the 'Our Data Bodies' project which examined data-based discrimination in the United States via interviews with marginalized citizens in Charlotte, Detroit, and Los Angeles. In their 'Reclaiming Our Data' report,[62] participants reflected on their experiences of data extraction and identified numerous concerns. There was the view that 'my data doesn't represent who I am', which reflects the sense that systems are dehumanizing, treating people as mere data points, as opposed to actual people. Then there was the feeling that 'systems are working against, rather than for, them', not least because technologies are often too intrusive, with people having little choice but to give up information in the knowledge that value from the data collected will benefit technology providers, not those to whom the data refer. In sum, contributors simply did not trust data-driven systems and did not believe they were designed or used to improve their lives. Rather, they viewed data-driven technologies as alienating and predatory, as perpetuating structural inequalities by routinely holding them back, diverting them away from resources they were entitled to, and ultimately exacting an emotional toll by making them feel 'uneasy, frustrated, overwhelmed, frightened, and angered'.[63] These communities are also the least likely to have the education, skills, tools, and data literacy to protect themselves from data extraction and to challenge the outcomes of profiling and sorting. And even if they can make sense of and create counterpoints to data-driven discrimination, they do not necessarily have a public voice or the political skill to take on a well-resourced, powerful opponent. For these communities, then, data extraction is a profound issue.

As this discussion highlights, there are significant questions concerning fairness, equity, and justice in our digital society. Some groups are much more likely than others to experience the pernicious effects of acceleration and data extraction, while at the same time being much less able to combat them and practise slow computing. And because these issues are structural and not easily fixed by an individual, they require a collective ethics of digital care to address. This requires people working together to campaign for change by exerting pressure on the owners and controllers of digital systems to make their systems more inclusive and empowering. For everyone to experience the joy of slow computing and enjoy

a balanced digital society, we must enact an ethics of digital care
that permits people to acquire control over their digital lives.

7

Towards a More Balanced Digital Society

Sunday. 10pm. The week is over. You've done well. You've managed to move away from frantic messaging and updating without giving up being online. You've taken note of your data trails and adjusted your practices along with others to slow down and evade excessive data extraction. Life isn't perfect all of a sudden, but you've been taking part in a reassessment of your digital life and you've some more control now. The weekend has been relaxing without being constantly tethered to the internet. Tomorrow is Monday and the day will start again at 6.30am. The question is whether the working week will be as crazily busy and stressful as usual, or whether it'll be possible to slow it down and be calmer, more focused and balanced. You think about checking your email and social media accounts, but decide it can wait until tomorrow. There's no point ruining a night's sleep by worrying about the contents of a message or the state of the world according to Facebook. In fact, you think you might give social media a break for a few days. The world is not going to end because you only log in a couple of times a week. You turn off the light and quickly fall into a deep sleep.

This doesn't sound too bad, does it? You're still feeling the joy of computing, but more on your terms. You've achieved some kind of balance between acceleration and data extraction and getting on with your everyday life. You've found some harmony between work and home. You're living a slow computing life. You're no longer a slave to your devices; you've more freedom from the digital leash. You're less harried, stressed and distracted. You're not careening through life, but doing more of the things you want, when you want to, and you're not as beholden to the demands of others. You feel like you have more time for yourself – for rest, leisure, personal interests and goals, and contemplation. You're doing fewer things, but you're enjoying what you do more. In fact, you feel like you've become more productive and your work has improved. It seems like you've got more control over your digital footprints and shadows. You're still using services, but you have a hand in deciding what data are gathered. You have a better sense of your rights and an understanding about how to challenge malpractice.

Of course, you're not fully in charge: after all, you still have responsibilities and obligations to others, you're still bound to the interests of institutions and companies, and there's only so much tinkering you can do with digital devices and infrastructures. And practising slow computing has meant accepting a degree of inconvenience in how you go about some things and being willing to make sacrifices so as to practise an ethics of digital care for others. But on the whole it is worth it. The benefits outweigh the negatives. Life is always a series of compromises and slow computing lets you come out top.

You also know that you will need to continually work at maintaining a slow computing life. Our digital devices, the platforms and infrastructures we use, are thoroughly integrated into our everyday lives. We are beholden to them in many cases, unable to complete tasks without using them. Moreover, there will be pressures to speed up again – from family members who always want to be doing things; from friends who like to meet up on the fly; from employers and clients who want you to be available 24/7; and from service providers who want to nudge you back into constant responsiveness. The compulsive and addictive nature of some platforms might also be difficult to resist. There will be a continual struggle with devices, apps, and services to minimize data extraction – constantly having to review terms and conditions, tweaking settings, update privacy enhancement tools, practise obfuscation, and campaign for the adoption of fair information practice principles and privacy-by-design.

There might be an ongoing battle to convince family, friends, and employers that slow computing doesn't just work for you, but also for them. You won't be so tired and grumpy, you'll have more fulfilling collaborations and relationships with colleagues and clients, you'll be more productive and have better ideas, and you'll be taking fewer days off ill. Of course, you are not alone in these struggles. Slow computing is a collective as well as individual effort. Our argument is not that slow computing entails halting or dismantling a data-driven society or economy, nor is it about taking the joy out of computing. Rather, it is about seeking balanced responses, mutual consensus, and win-win situations. We can all benefit from embracing slow ideas and ideals. A balanced digital society is possible and desirable. In this final chapter, we reiterate the suite of individual and collective tactics that we can adopt to enact slow computing and discuss the challenge of encountering and trying to overcome persistent obstacles.

Downshifting

Given the continued acceleration of everyday life and its effects on pace, tempo, scheduling, and timeshifting, it seems plausible that in the coming years everyone will want the option to downshift. Downshifting is usually associated with a radical change in lifestyle; for example, giving up a stressful job in a busy city to move to a more sedate rural location and work in a less pressured environment.[1] Moving from work to retirement is the most common form of downshifting, though that is not available to everyone, with many people having little or no pension. We are using downshifting here as a synonym with slow computing: as the ability to slow down and scale back, even for a short period of time. It might be aspirational to be able to do a conventional downshift – to wipe the slate clean and adopt a totally new lifestyle – but realistically, given various obligations and resources, that's not an option for most. The best we can hope for is a relative downshifting within an existing context.

Downshifting is an explicit expression of an ethics of care. It is an attempt to protect our own wellbeing, and also that of others, by gaining time sovereignty: that is, control and autonomy over the temporality of our lives. As we have detailed, there are a number of ways in which we can stake a claim to time sovereignty. Some of them are individual practices, others are more collective in nature. We think it is useful here to pull out and list the various tactics – some of which are individual, some collective – that we discussed in other chapters to provide a touchstone you can quickly refer back to. It is not an exhaustive list and we're sure you can probably think of other tactics to add, some of which you might already employ (for a fuller discussion of these tactics, see Chapters 4 and 5):

- practise structured rest and work;
- block out working time, minimizing digital distractions during these periods;
- identify aspects of life that you can step away from with little consequence;
- refuse to engage with colleagues and clients outside work hours, or at least only attend to critical cases;
- have separate devices for home and work;
- keep home phones free of work-related apps;
- only use the work phone during work hours;
- create a set of non-networked workspaces where workers have minimal distraction;
- valorize quality of work, workplace satisfaction, and worker health;
- restrict the times that work emails can be sent, avoiding non-work hours;
- only send circular work emails at set times, rather than continually across a day;
- only send circular emails to those they really concern rather than everybody;
- avoid last minute altering of schedules;
- enable workers to have the right to disconnect;
- respect the sanctity of public holidays and individual vacations;
- cater for different styles of learning and working;
- slow down the pace and tempo of work, play, and home and social life;
- reduce your social circle to those friends you really want to keep connected to;
- change how you engage with digital life, disconnecting when possible:

- wait until after breakfast before reaching for your smartphone;
- read a book on the train or bus rather than using a device;
- turn off notifications at meal and break times;
- check your email or social media at set times and answer in intense bursts;
- prioritize what needs attention and park what can be ignored;
- only engage in social media threads if it is truly worth spending time on;
- switch off home Wi-Fi routers in the evening or at weekends;
- leave phones/tablets in 'airplane mode' when you want to be disconnected;
- do not use a digital device within an hour of bedtime and leave devices outside your bedroom overnight or turn them off;
- delete apps that are too compulsive and take up too much time for little reward.
- deliberately engage in analogue practices:
  - organize device-free meals at set times;
  - revert to using clock time to arrange meetings;
  - use a paper diary and read newspapers, rather than online calendars and news;
  - try to avoid attending meetings and social events arranged on the fly;
  - wait in line rather than using self-scanning services;
  - use paper timetables rather than a smartphone app;
  - do your shopping in local stores rather than online.

These tactics take the need for slowness seriously. They enable downshifting. They create time by countering acceleration and freeing us from network time. They provide a degree of individual and collective time sovereignty. On paper, they seem relatively straightforward. Common sense, even. Yet, as we've discussed, they can be tricky to achieve in practice. One way to start, we think, is to adopt a handful of them and work from there, adding more as your lifestyle changes. And keep at them. Ideally, you also want to 'switch off smart': that is, use these tactics but in a way that lets others know what you've done and why. After all, we're trying to reduce stress, not incite it in others.

Seeking anonymity

We are rapidly moving to a situation in which just about all aspects of everyday lives – what we do, how we perform, where we go, who we meet, what we say, what we consume – will be recorded. George Orwell famously prophesized in his dystopian novel *1984* that there would be cameras constantly watching us everywhere, including in our own homes. What he did not anticipate is that we would willingly be buying and using devices that actively gather data about us,[2] nor that we would openly share our thoughts, opinions, photos, videos, activity, movements, and so on. Not only that, but as comic Keith Lowell Jensen joked, for many our biggest fear is not necessarily data capture, but that nobody is watching or acting on the data.[3] As we have discussed throughout this book, this is not a position shared by everybody. Many of us want to protect our privacy. We want to be sheltered from profiling, social sorting, and targeting. We want data-driven systems to treat us fairly and to work for us, not against us. We want anonymity, and not just for 15 minutes.

We desire an ethics of digital care that does not derail the state from performing its duties or prevent companies from turning a profit, but at the same time acts in our interests and curbs excessive and exploitative practices. This ethics of digital care provides us with a degree of data sovereignty – that is, some power and self-determination with respect to what data are captured about us, what happens to that data, and what the data are used for. Throughout the book we have documented a number of individual and collective ways in which we can stake a claim to data sovereignty. Here, we list some of those tactics to act as a quick reference for practical and political ideas, with a more detailed discussion in Chapters 4 and 5. Again, you can probably think of alternative tactics to add to the list:

- curate devices and services to turn on privacy settings and reduce exposure;
- use open source software and operating systems;

- turn-off Wi-Fi and Bluetooth on smartphones and tablets when not needed;
- use private (incognito) browser sessions and actively manage/delete cookies and browser history;
- use privacy enhancement tools, such as Https Everywhere, Privacy Badger, Ghostery, and ad and pop-up blockers;
- install obfuscation plugins, such as Do Not Track and AdNauseam;
- use a VPN service or a Tor browser;
- employ encrypted apps for messaging and cloud services;
- use USB 'data condoms' and camera covers;
- ensure up-to-date virus and malware tools are active;
- use a non-tracking search engine, such as DuckDuckGo;
- make up pseudonyms or false identities to sign into temporary services such as free Wi-Fi;
- use different usernames, profiles, and passwords across platforms;
- check the provenance of websites, using internet safety services;
- leave services, close down accounts, and remove apps that do not provide sufficient protection from harms;
- limit the services you register with and only subscribe to those you use repeatedly;
- be prepared to exercise rights to notice, consent, choice, and access, and to challenge data controllers and processors;
- actively curate digital footprints – carefully selecting the information you willingly share with social media and other apps;
- avoid quizzes or games that ask for personal information;
- encourage market-based solutions, self-regulation, and alternative business models;
- promote privacy as a means to achieve competitive advantage;
- challenge the orthodoxy at work to change data extraction and management practices;
- campaign for privacy and data protection rights and regulations;
- hold companies and the state to account with respect to regulations and laws;
- encourage privacy-by-design as standard and best practice;
- support the employment of data protection officers, privacy teams, and external privacy advisory boards;
- put in place education and training programmes to improve knowledge of issues and rights and to provide practical skills;

- take an active role in creating and promoting civic tech, community-run, platform co-ops, hackathons, and data dives;
- take an active role in creating and promoting open source software, tools, and resources;
- support organizations which are actively tackling data extraction and data justice issues such as Privacy International, ACLU, and EFF.

All these tactics seek to provide an ethics of digital care with respect to data extraction. None of them on their own will provide a strong degree of data sovereignty, but when several are used together they can provide a modicum of control. To do so, they need active intervention on an ongoing basis. Indeed, doing any of these tactics just once will limit the timespan of any effect. Given the value that can be gained from data, many of these tactics will be resisted. Surveillance capitalism and the attention economy will keep mutating to find new ways to capture and monetize data. New tactics will be needed to effectively counter any pernicious effects. As time rolls on, then, you will need to try and keep yourself abreast of new developments and react accordingly. This obviously involves some overhead in terms of time and effort, though the investment should hopefully be rewarded through the maintenance of your data sovereignty. And any work you have done collectively will hopefully provide a trickledown of care for others.

Persistent obstacles

Implementing these tactics and living a slow computing life is not going to be easy, as we've stressed throughout this book. There are a number of persistent obstacles that make tackling acceleration and data extraction difficult. Powerful vested interests would prefer to maintain the status quo. Exploiting digital life is profitable. Digital data and ubiquitous computation enable value to be extracted, create new markets, make existing services and processes more efficient, and increase productivity and competitiveness. Digital life makes managing society more effective: it provides deeper, wider, and more timely data about citizens; enables these data to be linked together to gain more insights; and creates scope for more real-time control of operational systems. Digital life offers people real, tangible benefits – instant communication and access to information, increased choice and an easy hunt for competitive prices, a whole raft of useful tools and devices, widened access to entertainment, and flexibility and convenience in organizing everyday life. Companies and states are reluctant to change a situation that is working well for them, and people want to continue to experience the joy of computing.

In fact, companies actively work to not only reproduce the current system but also deepen our reliance on digital technologies and block any slow computing efforts. They move towards discontinuing the production of analogue or non-connected digital products so that only 'smart', networked devices are available.[4] They roll out new products that demand frequent interaction and/or enable them to gather more data. They employ neuroscientists and behavioural specialists to find ways to make their products more compulsive and addictive. They tweak goods and services to reduce what scope users have to slow down or dance around platforms and devices. Software updates override adjustments to settings, install new applications, or tweak the arrangement of individual apps/ devices to prompt certain types of behaviour. Employers actively block the forming of unions or seek to undermine their efforts, and they continue towards using precarious labour. And they spend millions on lobbying politicians and conducting PR campaigns to block regulatory change or to deregulate or open new markets.[5]

Much as any one individual might like to practise slow computing and resist the dominant tendencies, some firms appear to be well aware of the need to remain one step ahead of resistance. States compel citizens to engage with them through digital systems and to share data with them, they add digital control to their operations and roll out new digital services, and they seek to persuade the public of the rationale for these moves using discourses of convenience, efficiency, safety, security, combating fraud, and so on. Some of these systems (such as smart city technologies) act passively, extracting data without us even necessarily knowing it is happening. And we can experience peer pressure from family and friends, as well as colleagues and clients, to always be connected and responsive.

These conditions combine to establish persistent obstacles that need to be continually overcome. Fighting against these moves and pressures is hard work. It can be exhausting, even if you have the social standing and requisite skills to do so. It might seem like a very tall mountain to climb if you do not. As we discussed in the last chapter, practising slow computing is more difficult for some than for others. Parents and carers typically have less autonomy over their time due to obligations to others. People who hold insecure jobs are more tied to a digital leash. Those that do piecework or work in highly time-controlled environments have little opportunity to slow down. Workers in the gig economy necessarily have their 'dead time' colonized by work. Poorer and more marginalized people are much more closely monitored by the state, and are much more likely to be negatively affected by profiling and social sorting, and have little opportunity to opt out. People of colour and ethnic minorities are more likely to be subject to inherent biases in the value structures within systems and to experience racial profiling and institutional racism. Children and adults with development issues lack the ability to exert agency with respect to their time and data sovereignty. Disadvantaged communities are least likely to have the education, skills, tools, and literacy to practise slow computing. From this perspective, slowing down and disconnecting are a privilege, not an expectation or right.

What this discussion highlights is that there are a number of constraints on slow computing, which results in a maximum viable threshold of practices: that is, a point at which slow computing cannot continue because doing so would require overcoming seemingly insurmountable corporate, governmental, institutional, or social barriers. Slow computing – in terms of fully gaining time and data sovereignty – is to a large extent a utopian and unrealizable state of affairs.

Nevertheless, there is still room for partial successes. Persistent obstacles, after all, are there to be challenged and overcome. Sure, it might be difficult to make headway against them, but it is not impossible to change some of them in ways that will be beneficial. New laws in France and Italy concerning the right to disconnect, the implementation of the GDPR in the EU, the creation of new privacy tools, and the success of open source software, all show that progress can be made. There are always opportunities to practise slow computing and to challenge persistent obstacles. And this is what we need to strive for. We need to accept that slow computing will always produce a blend of realities, in some cases resulting in an 'accelerated deviation' from optimal slow computing (when there is no slowing down, but significant 'data dancing'), in others yielding an 'extraction deviation' (when there is significant slowing down, but little evasion of extraction). But we should also not give up on creating the conditions that enable slow computing to happen, not just for us, but also for others who might not have the same opportunities, and for future generations. This is why practising an ethics of digital care and collective slow computing are so important. They provide both the moral imperatives for trying to persuade others of the benefits to society of slow computing and the collective means to try and create momentum towards change. If we work to slow down together then we can move towards a slow computing world and establish more balanced digital lives.

Thinking about slow computing

So far in this closing chapter we have reiterated the tactics, practices and political moves that you can adopt to try and create a slow computing lifestyle. Throughout the book we have also provided conceptual ideas and tools for thinking about and making sense of the effects of acceleration and data extraction, including normative notions designed to enable an evaluation of what should be rather than simply what is. Alongside developing the idea of slow computing itself and an ethics of digital care, we have drawn upon and elaborated the conceptual ideas of time sovereignty, data sovereignty, ubiquitous computing, datafication, surveillance capitalism, dataveillance, digital leash, data dancing, data-mentality, digital detox, switch off smart, privacy-by-design, the right to disconnect, the hesitant present, network time, responsibilization, data determinism, anticipatory governance, the hook cycle, decentralized web, data justice, ethics of forgetting, ethics of the future, future present, present future, and others.

Some of these concepts are relatively straightforward, being quite narrow and tight in how they detail and explain an idea (for example, data dancing), others are much more extensive in their scope and their rationale and logic (for example, ethics of digital care and data justice). Indeed, these might be varied and contested, with different people setting out alternative formulations of ethical and justice thinking. Rather than unpack these concepts, detailing in depth the facets and logics of a concept and the wider intellectual traditions and debates in which they sit, we have sought to present them in an accessible way that highlights their utility in making sense of an issue. Our aim has been to use them to help you understand how and why the digital revolution has created issues of time and data sovereignty, explain why these issues matter and need redress, and equip you to tackle them personally and collectively, rather than to produce a more academic treatise.

However, if you have found these conceptual ideas useful, and you are interested in understanding slow computing further, then we'd encourage you to engage with the articles and books cited in the endnotes to help develop your thinking and active responses, perhaps even extending and deepening the ideas, or adapting or developing new ones to critique the nature of our digital lives. In our view, the further development of the conceptual ideas and tools that underpin slow computing would be a very welcome and worthy endeavour. Indeed, it would be a central tactic of pursuing slow computing from an intellectual perspective, providing evidence that can shape policy.

The joy of slow computing

We have reached the end of our journey. Hopefully we have persuaded you of the benefits of slow computing. That it is possible to experience the joy of computing – to continue to use digital technologies for all kinds of productive purposes – while also reining in some of the more problematic aspects of living digital lives. That we can be enmeshed in a digital world yet retain more time and data sovereignty. That society and economy will be more balanced and enhanced by adopting slow computing principles and its ethics of digital care. That not only individuals and communities will benefit, but also companies and government.

To help you transition to a slow computing lifestyle we have documented a number of practical and political tactics that you can adopt. We are not advocating that you take up all of them, but that you experiment to find which ones are achievable and work best for you in terms of improving your digital life. Your approach is best guided by reflecting on your situation and undertaking an audit strategy for determining the various ways you are ensnared in acceleration and data extraction and how you might personally go about disentangling yourself, as we detailed in Chapter 4.

It is important to note that some of the tactics you use might have an immediate effect, but others might need longer to take hold. You have to give slow computing a bit of time to pay back on the investment. Slow computing, in the end, is *slow*; it takes time because it's supposed to take time. And, as we have stressed throughout the book, you need to continually work at a personal and collective level to maintain slow computing in the face of persistent obstacles and pressure. It's also important to practise it with other forms of slow living – after all, digital technologies are not solely responsible for people feeling harried and rushed, and data about us are still generated through non-digital means. It will not always be an easy path to follow, but the rewards are worth it.

All that is left for us is to wish you every success in creating balanced digital lives and experiencing the joy of slow computing.

Coda: Slow Computing During a Pandemic

The coronavirus pandemic started to sweep across the globe just as this book was going to press. All aspects of daily life changed once delay, and then containment measures, were put in place. Initially, closing down most workplaces and schools and restricting movement seemingly created new scope for people to practise slow computing. Rather than dashing here and there, trying to cope with a crowded diary and too many tasks, those people not on the frontline would be static and confined to the home. Life would become stationary, routines broken, busyness reduced, and work-life balance restored. However, the scope for pursuing slow computing is now in question like never before.

In many ways our lives have become even more digitally-mediated. In our own cases, at very short notice we had to pivot our teaching from face-to-face contact on our university campus into virtual classes. New knowledge and skills had to be acquired about new pedagogies and platforms (Teams, Skype, Zoom, Moodle, etc). Classes and meetings were to be conducted from home. Social interactions with family and friends shifted to video calls, WhatsApp and Facebook. Information was elicited through social media and news sites. Streaming services replaced out-of-home social activities. Our time was still fragmented and interleaved, and rather than our sense of stress being lowered, it was heightened by the sense of isolation and the fear and anxiety expressed through our media channels. We thus tried to follow our own slow computing advice by limiting the use of social media and making sure to do non-digitally mediated activities: exercise, cooking, gardening, reading, playing traditional games.

We're fortunate. For some of our colleagues (and also our students), the new digital realities of working at home have posed acute challenges. Many were left looking after bored, cooped-up children who needed home schooling, play and reassurance. They've had to cope with family-wide fights over who will use the computer. New skills have been acquired to access and install software and work out how to use new services. Some have quite limited access to broadband internet. Other workers have not been allowed to self-isolate due to the nature of their job, performing essential work. In many cases, this work has intensified due to increased demand or the stress of trying to deliver it in difficult circumstances. At the same time, many of these essential workers are trying to deal with organizing childcare or other care duties when schools and crèches are closed and services limited. And in many jurisdictions they are doing this work with little protection against infection or access to needed health insurance. Others still have found themselves out of work at short notice and scrambling to negotiate government websites to access welfare and unemployment benefits.

In addition, new social and technological arrangements that amplify surveillance and data extraction practices have started to emerge in an effort to halt the spread of the pandemic virus. Led by governments and companies, these technologies have been rolled out for five primary purposes: (1) quarantine enforcement/movement permission (knowing people are where they should be, either enforcing home isolation for those infected or close contacts, or enabling approved movement for those not infected); (2) contact tracing (knowing whose path people have crossed); (3) pattern and flow modelling (knowing the distribution of the disease and its spread and how many people passed through places); (4) social distancing and movement monitoring (knowing if people are adhering to recommended safe distances and to circulation restrictions); and (5) symptom tracking (knowing whether the population are experiencing any symptoms of the disease).[1]

Numerous digital technologies are employed to perform these tasks, including smartphone apps, facial recognition and thermal cameras, biometric wearables, smart helmets, drones, and predictive analytics.[2] For example, citizens in some parts of China have been required to install an app on their phone and then scan QR codes when accessing public spaces (e.g., shopping malls, office buildings, communal residences, metro systems) to verify their infection status and permission to enter.[3] The Polish government introduced a home quarantine app that required people in isolation to take a geo-located selfie of themselves within 20 minutes of receiving an SMS or risk a visit from the police.[4] Israel repurposed its advanced digital monitoring tools normally used for counterterrorism to track the movement of phones of all coronavirus carriers in the 14 days prior to testing positive in order to trace close contacts.[5] As of mid-April, 28 countries had produced contact tracing apps that use Bluetooth to detect and store the details of nearby phones and contacts them if someone who had been near them tested positive, and another 11 were planning to launch imminently.[6] Other states have utilised technologies designed to measure biometric information. For example, hand-held thermal cameras have been used in a number of countries, some mounted on drones, to screen movement in public space.[7]

Technology companies have offered, or have actively undertaken, to repurpose their platforms and utilise the data they hold about people as a means to help tackle the virus. Most notably, Apple and Google, who provide operating systems for iOS and Android smartphones, are developing solutions to aid contact tracing.[8] In Germany, Deutsche Telekom are providing aggregated, anonymized information to the government on people's movements; likewise Telecom Italia, Vodafone and WindTre are doing the same in Italy.[9] Unacast, a location-based data broker, is using GPS data harvested from apps installed on smartphones to determine if social distancing is taking place,[10] with several other companies offering similar locational and movement analysis. Experian, a large global data broker and credit scoring company, has announced it will be combing through its 300 million consumer profiles to identify those likely to be most impacted by the pandemic and offering the information to 'essential organizations', including health care providers, federal agencies and NGOs.[11] Some of the most problematic aspects of surveillance capitalism have been repurposed by the state, further legitimating and cementing their practices.

Beyond society-wide surveillance to combat the pandemic, some companies have rushed to implement their own versions of these technological solutions, for example scanning the temperature of workers or deploying their own contact tracing systems. These are likely to become more common as restrictions are lifted, and their use might become a mandatory condition of entering workplaces. In addition, many have adopted remote work surveillance systems so they can monitor the activity and productivity of their employees working at home, including recording keystrokes, how many emails are sent and their contents, and what employees are printing, or seeking constant status updates or that work is always undertaken while a video call is live.[12] These companies argue that they are trying to ensure that their workers are not taking unfair advantage of flexible work arrangements, or are not leaking confidential information. They take no account of workers trying to cope with the change in workplace environment which may not be conducive to work due to increased care duties, living in a shared space, or having poor or no broadband. Or workers have to learn new systems and procedures at short notice, or do not necessarily have the technical competence to perform any IT services needed to set up and maintain home-based work.

Some citizens will no doubt embrace surveillance technologies regardless of potential deleterious effects in the hope they will help to limit the spread of the virus and thereby save lives. Others might argue that companies should be able to know if their employees are performing the work they are paid to do. An underlying problem, however, stems from the track record of digital technology providers and governments in handling, protecting and extracting value from data. It seems logical to expect that data on movements, contacts or health will have value beyond the current public health crisis and they will be repurposed in some way that is not necessarily beneficial to citizens.[13] There are legitimate concerns as to whether public health and workplace surveillance systems will be turned off after the crisis or whether they will become a normal part of a new surveillance regime, as was the case with systems adopted after 9/11. Without embracing data sovereignty, privacy, civil liberties, workers' rights, citizenship and democracy are under renewed threat.[14]

In this regard it is significant that civil liberties organizations have set out ethical principles designed to protect privacy and rights, while acknowledging the potential utility of digital tools to tackle the virus. The key argument is that we should strive to ensure both civil liberties *and* public health, rather than simply trading the former for the latter. For example, the Electronic Frontier Foundation,[15]

American Civil Liberties Union,[16] the Ada Lovelace Institute,[17] and the European Data Protection Board[18] have demanded that:

- data collection and use must be based on science and need;
- the tech must be transparent in aims, intent, and workings;
- the tech and wider initiative must have an expiration date;
- a privacy-by-design approach with anonymization, strong encryption and access controls should be utilized;
- tools should be opt-in with consent sought, with very clear explanations of the benefits of opting in, operation and lifespan;
- the specification and user requirements, a data protection/privacy impact assessment, and the source code for state-sanctioned coronavirus surveillance should be published;
- data cannot be shared beyond the initiative or repurposed or monetized;
- no effort should be made to re-identify anonymous data;
- the tech and wider initiative must have proper oversight of use, be accountable for actions, have a firm legislative basis, and possess due process to challenge mis-use.

In other words, the tools must only be used when deemed necessary by public health experts for the purpose of containing and delaying the spread of the virus and their use should be discontinued once the crisis is over. We would add that we must also be vigilant to any potential control creep; that is, the risk that apps designed to limit movement based on health status will continue to be used and their criteria extended.

The temporal and organizational aspects of tackling the coronavirus pandemic raise other questions about the ethics of digital care. How do we ensure wellbeing and protect our civil rights while responding rapidly to an emerging crisis? How can we find a balance between the interests of public health and the economy and our own self-care? We don't have ready answers to these questions; formulating individual and collective interventions for slow computing within such a context is not straightforward. We are all now dealing with radically different circumstances. But an obvious conclusion to draw about the crisis response hitherto is that employers and employees need to define and deliver an ethics of digital care. For sure, some managers will have pursued admirable practices: facilitating flexibility and accommodating workers with respect to workload, hours, and deliverables. Others might have been trying to maintain a business-as-usual stance, thereby elevating stress levels on employees or colleagues.

At the same time, the ethics of digital care concerns those people struggling with non-reciprocal care duties, experiencing the ill-effects of social isolation, or becoming obsessed with media stories that elevate anxiety and place a strain on mental health. New pressures have been placed on women, particularly working mothers, who find their duties increasing and societal supports shrinking. And for the working poor, the ethics of digital care are given new meaning when they find themselves negotiating online government sites to access support, or working essential frontline jobs in retail, public transit, care, cleaning and so on with less protection while also subjected to regimes of digitally-mediated oversight. Practicing slow computing in such situations is not easy when one is bound within digital chains and societal expectation.

No matter what society emerges on the other side of this crisis, digital technologies are still going to be a fundamental part of our everyday lives. Indeed, the crisis might lead to elevated levels of remote working, virtual meetings, digitally-mediated interactions, and online consumption; after all, the response has demonstrated that these can adequately supplement or replace some existing work and social practices. As such, smartphones, personal computers, smart city systems, social media, streaming services, online consumption, games, e-governance, and so on, will continue to saturate and configure our time and extract and utilise our data. Enhanced surveillance and dataveillance practices might remain in place, meaning that it will become ever more necessary to try and protect oneself from data extraction, ensure privacy, and push back against new pernicious powers. More than ever, an ethics of our digital future is required; at issue is a duty of care to imagine and create a society that enables us to practice slow computing during a fast response to a crisis and subsequent recovery. Individual and collective slow computing will remain necessary if we are to experience the joy of computing and enjoy balanced digital lives.

Notes

Chapter 1

1   Deloitte (2017a) *The Dawn of the Next Era in Mobile. 2017 Global Mobile Consumer Survey: US edition.* https://www2. deloitte.com/content/dam/Deloitte/us/Documents/ technology-media-telecommunications/us-tmt-2017-global-mobile-consumer-survey-executive-summary.pdf. An IDC report from 2013 reported 62% of Americans reach for their phone immediately after waking, with a further 17% within 15 minutes. For 18–24 year olds 74% reach immediately for their phone and 89% within 15 minutes. IDC (2013) *Always Connected.* https://www.nu.nl/files/IDC-Facebook%20 Always%20Connected%20%281%29.pdf

2   39% of 18–29 year olds and 35% of 30–49 year olds report being almost constantly online. http://www.pewresearch.org/ fact-tank/2018/03/14/about-a-quarter-of-americans-report-going-online-almost-constantly/. Also https://www.asurion. com/about/press-releases/americans-dont-want-to-unplug-from-phones-while-on-vacation-despite-latest-digital-detox-trend/.

3   Deloitte (2017a). Over 90% of people aged 18–44 possess a smartphone.

4   In a 2015 report, 71% of Americans slept with their smartphone on a nightstand, in their bed, or, for 3% of people, in their hand. http://fortune.com/2015/06/29/sleep-banks-smartphones/.

5   Deloitte (2017b) *State of the Smart: Consumer and Business Usage Patterns. Global Mobile Consumer Survey 2017: UK cut.* https:// www.deloitte.co.uk/mobileuk2017/assets/img/download/ global-mobile-consumer-survey-2017_uk-cut.pdf.

6   Lustig, R. (2017) *The Hacking of the American Mind: The Science Behind the Corporate Takeover of Our Bodies and Brains.* Penguin, London; Eyal, N. (2014) *Hooked: How to Build Habit Forming Products.* Portfolio Books, New York.

7   Greenfield, A. (2006) *Everyware: The Dawning Age of Ubiquitous Computing*. New Riders, Boston.

8   Kitchin, R. and Dodge, M. (2011) *Code/Space: Software and Everyday Life*. MIT Press, Cambridge, MA.

9   SWNS (2017) Americans check their phones 80 times a day: study. *New York Post*, 8 Nov. https://nypost.com/2017/11/08/americans-check-their-phones-80-times-a-day-study/.

10   Dodge, M. and Kitchin, R. (2004) Flying through code/space: The real virtuality of air travel. *Environment and Planning A* 36(2): 195–211.

11   For example, an automatic number plate recognition (ANPR) camera captures the identity of every car on the road network, not a sample of them; Twitter captures every tweet posted by every user, not a sample of them. However, not all people moving in the city do so by car, and not everyone uses Twitter, hence the data are exhaustive to a system, but not the entire population.

12   Kitchin, R. (2014) *The Data Revolution: Big Data, Open Data, Data Infrastructures and Their Consequences*. Sage, London.

13   Lyon, D. (ed) (2003) *Surveillance as Social Sorting: Privacy, Risk and Digital Discrimination*. Routledge, London; Graham, S. D. N. (2005) Software-sorted geographies. *Progress in Human Geography* 29(5): 562–80; CIPPIC (2006) *On the Data Trail: How Detailed Information About You Gets into the Hands of Organizations with Whom You Have No Relationship. A Report on the Canadian Data Brokerage Industry*. The Canadian Internet Policy and Public Interest Clinic, Ottawa. http://www.cippic.ca/uploads/May1-06/DatabrokerReport.pdf.

14   Amoore, L. (2013) *The Politics of Possibility: Risk and Security Beyond Probability*. Duke University Press, Durham, NC.

15   Lane, J., Stodden, V., Bender, S. and Nissenbaum, H. (eds) (2014) *Privacy, Big Data and the Public Good*. Cambridge University Press, Cambridge.

16   Solove, D. J. (2006) A taxonomy of privacy. *University of Pennsylvania Law Review* 154(3): 477–560.

17   Baracos, S. and Nissenbaum, H. (2014) Big data's end run around anonymity and consent. In Lane, J., Stodden, V., Bender, S. and Nissenbaum, H. (eds) *Privacy, Big Data and the Public Good*. Cambridge University Press, Cambridge, pp 44–75; Crawford, K. and Schultz, J. (2014) Big data and due process: Toward a framework to redress predictive privacy harms. *Boston College Law Review* 55(1): 93–128.

18 Castells, M. (1996) *Rise of the Network Society*. Blackwell, Oxford.

19 Zuboff, S. (2019) *The Age of Surveillance Capitalism: The Fight for the Future at the New Frontier of Power*. Profile Books, New York.

20 Srnicek, N. (2016) *Platform Capitalism*. Polity Press, Cambridge.

21 See Ritzer, G. and Jurgenson, N. (2010) Production, consumption, prosumption. *Journal of Consumer Culture* 10(1): 13–36; Jarrett, K. (2016) *Feminism, Labour and Digital Media*. Routledge, London.

22 Schneider, N. (2015) The joy of slow computing. *New Republic*, 20 May. https://newrepublic.com/article/121832/pleasure-do-it-yourself-slow-computing.

23 Schneider, N. (2015) Slow computing. *America*, 21 May. https://www.americamagazine.org/content/all-things/slow-computing.

24 Honoré, C. (2005) *In Praise of Slowness: Challenging the Cult of Speed*. HarperCollins, New York.

25 Shojai, P. (2017) *The Art of Stopping Time*. Rodale Books, New York.

26 Craig, G. and Parkins, W. (2006) *Slow Living*. Berg, Oxford; Miele, M. and Murdoch, J. (2002) The practical aesthetics of traditional cuisines: Slow food in Tuscany. *Sociologica Ruralis* 42(4): 312–28.

27 Berg, M. and Seeber, B. (2017) *The Slow Professor: Challenging the Culture of Speed in the Academy*. University of Toronto Press, Toronto; Walker, M. B. (2016) *Slow Philosophy*. Bloomsbury, London; Mountz, A., Bonds, A., Mansfield, B., Loyd, J., Hyndman, J., Walton-Roberts, M., Basu, R., Whitson, R., Hawkins, R., Hamilton, T. and Curran, W. (2015) For slow scholarship: A feminist politics of resistance through collective action in the neoliberal university. *ACME: An International Journal for Critical Geographies* 14(4): 1235–59.

28 Datta, A. and Shaban, A. (2017) Slow: Towards a decelerated urbanism. In Datta, A. and Shaban, A. (eds) *Mega-Urbanization in the Global South: Fast Cities and New Urban Utopias of the Postcolonial State*. Routledge, London, pp 205–20. Cittaslow is an organization founded in 1999 dedicated to improving the quality of life in towns by slowing down its overall pace, see http://www.cittaslow.org/.

29 Fullagar, S. and Markwell, K. W. (2012) *Slow Tourism: Experiences and Mobilities*. Channel View Publications;

Clancy, M. (2016) *Slow Tourism, Food and Cities: Pace and the Search for the 'Good Life'*. Routledge, London.

30 Benjamin, R. (2019) *Race After Technology*. Polity Books, Cambridge, p 17.

31 Our conception of slow computing is rooted in a view that the relationship between digital technologies and time is not deterministic (that is, temporal effects are an inevitable outcome of technologies) but rather is contingent, relational, and contextual. That is, the temporal logic of the digital age is partly based on the social determination of time built into digital technologies and how these technologies are used in social context. In other words, society shapes the uptake and use of technologies, as much as these technologies shape society. There is nothing inevitable in the relationship. As Judy Wajcman puts it: 'there is a mutual shaping or coevolution of new technologies and temporal rhythms' (p 4). As such, time is shaped by technology but is not determined by it, and we can actively intervene in this relationship and reconfigure it – in our case, by slowing it back down. Wajcman, J. (2015) *Pressed for Time: The Acceleration of Life in Digital Capitalism*. Chicago, University of Chicago Press.

32 Woodruff, A., Augustin, S. and Foucault, B. (2007) Sabbath day home automation: 'It's like mixing technology and religion', *CHI '07: Proceedings of the SIGCHI Conference on Human Factors in Computing Systems*, pp 527–36.

33 Ems, L. (2015) Exploring ethnographic techniques for ICT non-use research: An Amish case study. *First Monday* 20(11). https://journals.uic.edu/ojs/index.php/fm/article/view/6312/5139.

34 Rheingold, H. (1999) Look who's talking. *Wired* 7(1), https://www.wired.com/1999/01/amish/.

35 Ems, L. (2015).

36 Ash, A. (nd) The best books on Slow Living recommended by Carl Honoré. Five Books. https://fivebooks.com/best-books/slow-living-carl-honore/.

Chapter 2

1 Eurofound (2016) *Working Time Developments in the 21st Century: Work Duration and Its Regulation in the EU*. Publications Office of the European Union, Luxembourg. https://www.eurofound.europa.eu/sites/default/files/ef_publication/field_ef_document/ef1573en.pdf.

2 Eurofound (2016).

3 Eurofound (2016) p 63.

4 Wajcman, J. (2008) Life in the fast lane? Towards a sociology of technology and time. *British Journal of Sociology* 59(1): 59–77.

5 Green, N. (2002) On the move: Technology, mobility, and the mediation of social time and space. *The Information Society* 18(4): 281–92.

6 Pang, A. S-K. (2016) *Rest: Why You Get More Done When You Work Less*. Basic Books, New York.

7 Pang, A. S-K. (2016) p 8.

8 Huffington, A. (2016) Foreword. In Pang, A. S-K. *Rest: Why You Get More Done When You Work Less*. Basic Books, New York, p xiii.

9 Crary, J. (2013) *24/7: Late Capitalism and the Ends of Sleep*. Verso, London.

10 Pang, A. S-K. (2016) p 24.

11 Pang, A. S-K. (2016).

12 Sutko, D. and de Souza e Silva, A. (2010) Location-aware mobile media and urban sociability. *New Media & Society* 13(5): 807–23.

13 Wajcman, J. (2008).

14 Kitchin, R. (2017) The realtimeness of smart cities. *Tecnoscienza* 8(2): 19–42.

15 Southerton, D. and Tomlinson, M. (2005) 'Pressed for time' – The differential impacts of a 'time squeeze'. *The Sociological Review* 53(2): 215–40; Wajcman, J. (2015).

16 Janelle, D. (1968) Central place development in a time-space framework. *Professional Geographer* 20(1): 5–10.

17 Crang, M. (2007) Speed = distance/time chronotopographies of action. In Hassan, R. and Purser, R. (eds) *24/7: Time and Temporality in the Network Society*. Stanford University Press, Stanford, CA, pp 62–88.

18 Wajcman, J. (2015).

19 Wajcman, J. (2015).

20 Harvey, D. (1989) *The Condition of Postmodernity: An Enquiry into the Origins of Cultural Change*. Blackwell, Oxford.

21 Janelle, D. (1968).

22 Giddens, A. (1984) *The Constitution of Society: Outline of the Theory of Structuration*. University of California Press, Berkeley, CA.

23 This process has been long noted. Indeed, Karl Marx (1857) noted that a key aspect of capitalism is its drive to create new

markets and enable the accumulation of capital by 'annihilating space by time'. Marx, K. (1857/1973) *Grundrisse: Foundations of the Critique of Political Economy*. Penguin Classics, London.

24 Rosa, H. (2015) *Social Acceleration: A New Theory of Modernity*. University of Columbia Press, New York. Rosa says 9.5km, but we've said 10km for sake of simplicity.

25 Leyshon, A. (1995) Annihilating space? The speed-up of communications. In Allen, J. and Hamnett, C. (eds) *A Shrinking World? Global Unevenness and Inequality*. Oxford University Press, Oxford, pp 11–54.

26 1932 Atlas of the Historical Geography of the United States. https://dsl.richmond.edu/historicalatlas/138/a/.

27 Galton, F. (1881) *Isochronic Passage Chart for Travellers*. Published by the Proceedings of the Royal Geographical Society, 1881. Available at: https://upload.wikimedia.org/wikipedia/commons/8/86/Isochronic_Passage_Chart_Francis_Galton_1881.jpg.

28 Novak, M. (2013) What international air travel was like in the 1930s. Gizmodo, 27 Nov. https://paleofuture.gizmodo.com/what-international-air-travel-was-like-in-the-1930s-1471258414.

29 Broadberry, S. and Burhop, C. (2009) Real wages and labour productivity in Britain and Germany, 1871–1938. *Journal of Economic History* 70(2): 400–27.

30 Leyshon, A. (1995).

31 AT&T Long Lines, Places and Routes (nd) Maps, Diagrams and Lists. http://long-lines.net/places-routes/.

32 Tuning in: Communications technologies historically have had broad appeal for consumers. *Wall Street Journal* 1998. Available at: http://www.karlhartig.com/chart/techhouse.pdf.

33 History of Television. Wikipedia. https://en.wikipedia.org/wiki/History_of_television.

34 Hart, J. A., Reed, R. R. and Bar, F. (1992) The building of the Internet: Implications for the future of broadband networks. *Telecommunications Policy* 16: 666–89; Salus, P. (1995) *Casting the Net: From Arpanet to Internet and beyond...* Addison Wesley, Reading, MA.

35 O'Neill, J. E. (1995) The role of ARPA in the development of the ARPANET, 1961–1972. *IEEE Annals of the History of Computing* 17(4): 76–81.

36 Dodge, M. and Kitchin, R. (2000) *Mapping Cyberspace*. Routledge, London.

37 Berners-Lee, T. (1999) *Weaving the Web: The Past, Present and Future of the World Wide Web by its Inventor.* Orion Business, New York.

38 Kitchin, R. (1998) *Cyberspace: The World in the Wires.* John Wiley and Sons, Chichester.

39 O'Reilly, T. (2009) What is Web 2.0? 30 Sept. https://www.oreilly.com/pub/a/web2/archive/what-is-web-20.html

40 Kitchin, R. and Dodge, M. (2011).

41 Holst, A. (2019) Number of smartphone users worldwide from 2016 to 2021 (in billions). Statista, 23 Oct. https://www.statista.com/statistics/330695/number-of-smartphone-users-worldwide/.

42 Statista Research Department (2019) Internet of Things (IoT) connected devices installed base worldwide from 2015 to 2025 (in billions). Statista, 9 Aug. https://www.statista.com/statistics/471264/iot-number-of-connected-devices-worldwide/.

43 Gershenfeld, N., Krikorian, R. and Cohen, D. (2004) The internet of things. *Scientific American*, October, 76–81.

44 Kitchin, R. and Dodge, M. (2011).

45 Mitchell, W. J. (1998) The new economy of presence. *Environment and Planning B: Planning and Design* 25: 20–21.

46 Urry, J. (2000) Mobile sociology. *Sociology* 51(1): 185–203.

47 Hassan, R. (2003) Network time and the new knowledge epoch. *Time and Society* 12(2/3): 225–41.

48 de Lange, M. (2018) From real-time city to asynchronicity: Exploring temporalities of smart city dashboards. In Lammes, S., Perkins, C., Gekker, A., Hind, S., Wilmott, C. and Evans, D. (eds) *Time for Mapping: Cartographic Temporalities.* Manchester University Press, Manchester, pp 238–55.

49 Kitchin, R. (2017).

50 Gerbaudo, P. (2012) *Tweets and the Streets: Social Media and Contemporary Activism.* Pluto Press, London; Farris, D. (2013) *Dissent and Revolution in the Digital Age.* IB Tauris, London.

51 Aiken, S. (2019) Inside digital resistance in Cypherpunk Harbour. *Medium*, 1 Oct. https://medium.com/crypto-punks/digital-resistance-security-privacy-tips-from-hong-kong-protesters-37ff9ef73129.

52 Crang, M. (2007).

53 Breathnach, P. (1998) Exploring the 'Celtic Tiger' Phenomenon: Causes and consequences of Ireland's economic miracle. *European Urban and Regional Studies* 5(4): 305–16.

54 Kitchin, R. and Bartley, B. (2007) Ireland in the twenty first century. In Bartley, B. and Kitchin, R. (eds) *Understanding Contemporary Ireland*. Pluto Press, London, pp 1–26.
55 Vizard, F. (2004) Building the information superhighway. *Popular Mechanics* 171(1): 29–33.
56 Breathnach, P. (2000) Globalisation, information technology, and the emergence of 'niche' transnational cities: The growth of the call centre sector in Dublin. *Geoforum* 31(4): 477–85.
57 Coletta, C., Heaphy, L. and Kitchin, R. (2019) From the accidental to articulated smart city: The creation and work of 'Smart Dublin'. *European Urban and Regional Studies* 26(4): 349–64.
58 Of course, it has had other effects as well, such as dramatic urban-regional restructuring, with a large growth in population, extensive suburbanization of housing and office/industrial premises, poly-centric and uneven development, congestion, and widening social divisions. With respect to the latter, not all boats rose by the same degree, with some groups and places being left behind.
59 Wajcman, J. (2015).
60 Gleick, J. (1999) *Faster: The Acceleration of Just About Everything*. Pantheon Books, New York.
61 Wajcman, J. (2008); Hassan, R. (2007) Network time. In: Hassan, R. and Purser, R. (eds) *24/7: Time and Temporality in the Network Society*. Stanford University Press, Stanford, CA, pp 37–61.
62 Research by John Robinson reported in Wajcman (2015) p 64.
63 Ash, A. (nd).
64 Purser, R. E. (2002) Contested presents: Critical perspectives on 'real-time' management. In Adam, B., Whipp, R. and Sabelis, I. (eds) *Making Time: Time in Modern Organizations*. Oxford University Press, Oxford, pp 155–67.
65 Hassan, R. (2007) p 55.
66 Purser, R. E. (2002).
67 Kitchin, R. and Dodge, M. (2011).
68 Leccardi, C. (2007) New temporal perspectives in the 'High-Speed Society'. In Hassan, R. and Purser, R. (eds) *24/7: Time and Temporality in the Network Society*. Stanford University Press, Stanford, CA, pp 25–36.

69   Bleecker, J. and Nova, N. (2009) *A Synchronicity: Design Fictions for Asynchronous Urban Computing*. Situated Technologies, New York.

70   Bleecker, J. and Nova, N. (2009).

71   de Lange, M. (2018).

72   Greenfield, A. (2013) *Against Smart Cities*. Do Publications, New York.

73   Kitchin, R., Cardullo, P. and di Feliciantonio, C. (2019) Citizenship, social justice and the right to the smart city. In Cardullo, P., di Feliciantonio, C. and Kitchin, R. (eds) *The Right to the Smart City*. Emerald, Bingley, pp 1–24.

74   Wajcman, J. (2015).

75   Wajcman, J. (2015).

76   Rosa, H. (2015).

77   Rosa, H. (2015) p 294.

78   Rosa, H. (2015) p 295.

79   Seligman, M. E. P., Railton, P., Baumeister, R. F. and Sripada, C. (2016) *Homo Prospectus*. Oxford University Press, Oxford.

80   Seligman et al (2016) p 195.

81   Seligman et al (2016) p 186.

82   Wajcman, J. (2015).

Chapter 3

1   Cheney-Lippold, J. (2017) *We Are Data: Algorithms and the Making of our Digital Selves.* New York University Press, New York.

2   Cheney-Lippold, J. (2017).

3   Coffee, P. (2018) IPG confirms $2.3 billion deal to acquire data marketing company Acxiom. *Adweek*, 2 July. https://www.adweek.com/agencies/ipg-confirms-2-3-billion-deal-to-acquire-data-marketing-company-acxiom/.

4   Downey, S. (2013) Acxiom's letting you see the data they have about you (kind of). *Abine*, 4 Sept. https://www.abine.com/blog/2013/acxioms-letting-you-see-data/.

5   Singer, N. (2012) You for sale: Mapping, and sharing, the consumer genome. *New York Times*, 17 June, http://www.nytimes.com/2012/06/17/technology/acxiom-the-quiet-giant-of-consumer-database-marketing.html.

6   Dencik, L., Hintz, A., Redden, J. and Warne, H. (2018) *Data Scores as Governance: Investigating uses of citizen scoring in public services*. Cardiff: Data Justice Lab. https://datajustice.files.

wordpress.com/2018/12/data-scores-as-governance-project-report2.pdf.

7 Hein, B. (2014) Uber's data-sucking Android app is dangerously close to malware. *Cult of Mac*, 26 Nov. http://www.cultofmac.com/304401/ubers-android-app-literally-malware/.

8 Dodge, M. and Kitchin, R. (2005) Codes of life: Identification codes and the machine-readable world. *Environment and Planning D: Society and Space* 23(6): 851–81.

9 Lyon, D. (2007) *Surveillance Studies: An Overview*. Polity, Cambridge.

10 Dodge, M. and Kitchin, R. (2007) 'Outlines of a world coming into existence': Pervasive computing and the ethics of forgetting. *Environment and Planning B* 34(3): 431–45.

11 Curran, D. (2018) Are you ready? Here is all the data Facebook and Google have on you. *The Guardian*, 30 Mar. https://www.theguardian.com/commentisfree/2018/mar/28/all-the-data-facebook-google-has-on-you-privacy; if you want to delete your Facebook account the *NY Times* provides a guide: Chen, B. X. (2018) How to delete Facebook. *New York Times*, 19 Dec. https://www.nytimes.com/2018/12/19/business/delete-facebook-account.html. You should note that they give you a grace period of 30 days so you can change your mind if needed. The entire deletion process may take up to 90 days to purge all backups of your data from the company's servers; it is not clear whether metadata and anonymized derived data are also deleted.

12 Google provides an option to download all of the data it stores about you. See http://google.com/takeout.

13 Facebook lets you download data it holds about you. See https://www.facebook.com/help/1701730696756992.

14 Mayer-Schoenberger, V. and Cukier, K. (2013) *Big Data. A Revolution That Will Transform How We Live, Work, and Think*. John Murray Publishers, London; van Dijck, J. (2014) Datafication, dataism and dataveillance: Big Data between scientific paradigm and ideology. *Surveillance & Society* 12(2): 197–208; Zuboff, S. (2019).

15 Zuboff, S. (2019).

16 Koops, B. J. (2011) Forgetting footprints, shunning shadows: A critical analysis of the 'right to be forgotten' in Big Data practice. *SCRIPTed* 8(3): 229–56.

17 Angwin, J. (2014) *Dragnet Nation*. St Martin's Press, New York; Kitchin, R. (2015) Spatial big data and the era

of continuous geosurveillance. *DIS Magazine.* https://dismagazine.com/issues/73066/rob-kitchin-spatial-big-data-and-geosurveillance/.

18 Gardham, M. (2015) Controversial face recognition software is being used by Police Scotland, the force confirms. *Herald Scotland*, 26 May. https://www.heraldscotland.com/news/13215304.Controversial_face_recognition_software_is_being_used_by_Police_Scotland__the_force_confirms/; Wellman, T. (2015) Facial recognition software moves from overseas wars to local police. *New York Times*, 12 Aug. https://www.nytimes.com/2015/08/13/us/facial-recognition-software-moves-from-overseas-wars-to-local-police.html.

19 Dodge, M. and Kitchin, R. (2007) The automatic management of drivers and driving spaces. *Geoforum* 38(2): 264–75.

20 Vincent, J. (2014) London's bins are tracking your smartphone. *The Independent*, 10 June. https://www.independent.co.uk/life-style/gadgets-and-tech/news/updated-londons-bins-are-tracking-your-smartphone-8754924.html.

21 Kopytoff, V. (2013) Stores sniff out smartphones to follow shoppers. *Technology Review*, 12 Nov. http://www.technologyreview.com/news/520811/stores-sniff-out-smartphones-to-follow-shoppers/; Henry, A. (2013) How retail stores track you using your smartphone (and how to stop it). *Lifehacker*, 19 July. http://lifehacker.com/how-retail-stores-track-you-using-your-smartphone-and-827512308.

22 Leszczynski, A. (2017) Geoprivacy. In Kitchin, R., Lauriault, T. and Wilson, M. (eds) *Understanding Spatial Media*. Sage, London, pp 235–44.

23 The *New York Times* digital ad sales were about $200m in 2016: https://www.nytimes.com/2017/07/27/business/new-york-times-company-2q-earnings.html.

24 Zuboff, S. (2019).

25 CIPPIC (2006).

26 Lee, E. (2019) Netflix stock tumbles as U.S. subscribers decrease after price increases. *New York Times*, 17 July. https://www.nytimes.com/2019/07/17/business/media/netflix-earnings-subscribers.html.

27 Carbonara, P. (2018) Walmart, Amazon top world's largest retail companies. *Forbes*, 6 June. https://www.forbes.com/sites/petercarbonara/2018/06/06/worlds-largest-retail-companies-2018/.

28 A 'digitally native' company is one that's born in the digital era and whose product is entirely digital and often available only online.

29 Sigala, M. (2005) Integrating customer relationship management in hotel operations: Managerial and operations implications. *International Journal of Hospitality Management* 24(3): 391–413.

30 Manyika, J., Chiu, M., Brown, B., Bughin, J., Dobbs, R., Roxburgh, C. and Hung Byers, A. (2011) *Big Data: The Next Frontier for Innovation, Competition, and Productivity*. McKinsey Global Institute, New York.

31 Elwood, S. and Leszczynski, A. (2011) Privacy reconsidered: New representations, data practices, and the geoweb. *Geoforum* 42: 6–15.

32 Martínez-Ballesté, A., Pérez-Martínez, P. A. and Solanas, A. (2013) The pursuit of citizens' privacy: A privacy-aware smart city is possible. *Communications Magazine, IEEE* 51(6): 136–141; Santucci, G. (2013) Privacy in the digital economy: Requiem or renaissance? *Privacy Surgeon.* http://www.privacysurgeon. org/blog/wp-content/uploads/2013/09/Privacy-in-the-Digital-Economy-final.pdf.

33 Solove, D. (2006).

34 Minelli, M., Chambers, M. and Dhiraj, A. (2013) *Big Data, Big Analytics*. Wiley, Hoboken, NJ.

35 European Data Protection Supervisor (2014) Privacy and competitiveness in the age of big data: The interplay between data protection, competition law and consumer protection in the Digital Economy. https://secure.edps.europa.eu/ EDPSWEB/webdav/shared/Documents/Consultation/ Opinions/2014/14-03-26_competitition_law_big_data_ EN.pdf.

36 Graves, J. (2015) An exploratory study of mobile application privacy policies. *Technology Science*, 30 Oct. https://techscience. org/a/2015103002/.

37 Statista (2015) Number of apps available in leading app stores as of July 2015. Statista. http://www.statista.com/ statistics/276623/number-of-apps-available-in-leading-app-stores/.

38 Zang, J., Dummit, K., Graves, J., Lisker, P., and Sweeney, L. (2015) Who knows what about me? A survey of behind the scenes personal data sharing to third parties by mobile apps. *Technology Science*, 30 Oct. https://techscience. org/a/2015103001/.

39 European Data Protection Supervisor (2014).

40 Solove, D. (2013) Privacy management and the consent dilemma. *Harvard Law Review* 126: 1880–1903.

41 Solove, D. (2013) p 1888.

42 Rubinstein, I. S. (2013) Big Data: The end of privacy or a new beginning? *International Data Privacy Law* 3(2): 74–87. https://academic.oup.com/idpl/article/3/2/74/709082.

43 Tene, O. and Polonetsky, J. (2012) Big Data for all: Privacy and user control in the age of analytics. *Social Science Research Network*. http://ssrn.com/abstract=2149364.

44 Crump, C. and Harwood, M. (2014) Invasion of the data snatchers: Big Data and the Internet of Things means the surveillance of everything. ACLU, 25 Mar. http://www.aclu.org/blog/speakeasy/invasion-data-snatchers-big-data-and-Internet-things-means-surveillance-everything.

45 Cypherpunk Manifesto, cited in Angwin, J. (2014).

46 Article 29 Data Protection Working Party (2014) *Opinion 8/2014 on the Recent Developments on the Internet of Things.* http://ec.europa.eu/justice/data-protection/article-29/documentation/opinion-recommendation/files/2014/wp223_en.pdf; European Data Protection Supervisor (2014).

47 Angwin, J. (2014). She also notes that in 2013 Krux Digital had identified 328 separate companies tracking visitors to the top 50 content websites (p 167).

48 Rambam, S. (2008) Privacy is dead, get over it. Presentation at the Last Hope conference, New York. http://www.youtube.com/watch?v=Vsxxsrn2Tfs; Rubenking, N.J. (2013) Privacy is dead. The NSA killed it. Now what? *PC Mag*, 10 Sept. http://www.pcmag.com/article2/0,2817,2424193,00.asp.

49 Raley, R. (2013) Dataveillance and countervailance. In Gitelman, L. (ed) *'Raw Data' is an Oxymoron*. MIT Press, Cambridge, MA, pp 121–46 (quote, p 126); Solove, D. J. (2007) 'I've got nothing to hide' and other misunderstandings of privacy. *Social Science Research Network*, http://ssrn.com/abstract=998565.

50 See Thatcher, J., O'Sullivan, D. and Mahmoudi, D. (2016) Data colonialism through accumulation by dispossession: New metaphors for daily data. *Environment and Planning D* 34(6): 990–1006.

51 Cavoukian, A. (2009) *Privacy by Design: A Primer*. http://www.privacybydesign.ca/content/uploads/2013/10/pbd-primer.pdf; Cohen, J. (2012) What is privacy for? *Social Science Research*

Network. http://papers.ssrn.com/sol3/papers.cfm?abstract_id=2175406 (quote, p 2).

52 Dougherty, C. (2015) Google photos mistakenly labels black people 'gorillas'. *New York Times*, 1 July. https://bits.blogs.nytimes.com/2015/07/01/google-photos-mistakenly-labels-black-people-gorillas/.

53 Noble, S. U. (2018) *Algorithms of Oppression: How Search Engines Reinforce Racism.* New York University Press, New York.

54 Harcourt, B.E. (2006) *Against Prediction: Profiling, Policing and Punishing in an Actuarial Age.* Chicago University Press, Chicago; Noble, S. U. (2018).

55 Noble, S. U. (2018).

56 Benjamin, R. (2019) p 11.

57 Benjamin, R. (2019); Eubanks, V. (2019).

58 Clarke, R. (1988) Information technology and dataveillance. *Communications of the ACM* 31(5): 498–512; Raley, R. (2013).

59 Graham, S. (2005) Software-sorted geographies. *Progress in Human Geography* 29(5): 562–80.

60 Minelli et al (2013).

61 Angwin, J. (2014); Clifford, S. (2012) Shopper alert: Price may drop for you alone. *New York Times*, 9 Aug. http://www.nytimes.com/2012/08/10/business/supermarkets-try-customizing-prices-for-shoppers.html.

62 Tene, O. and Polonetsky, J. (2012) p 17.

63 Dencik et al (2018).

64 Baracos and Nissenbaum (2014); Crawford and Schultz (2014).

65 Ramire, E. (2013) The privacy challenges of big data: A view from the lifeguard's chair. *Technology Policy Institute Aspen Forum*, 19 Aug. http://ftc.gov/speeches/ramirez/130819bigdataaspen.pdf.

66 Stroud, M. (2014) The minority report: Chicago's new police computer predicts crimes, but is it racist? *The Verge*, 19 Feb. http://www.theverge.com/2014/2/19/5419854/the-minority-report-this-computer-predicts-crime-but-is-it-racist.

67 Kitchin, R. and Dodge, M. (2011).

68 Cheney-Lippold, J. (2017) p 157.

69 The Cambridge Analytica Files. *The Guardian.* https://www.theguardian.com/news/series/cambridge-analytica-files; Pybus J. (2019) Trump, the first Facebook President: Why politicians need our data too. In Happer C., Hoskins A. and Merrin W. (eds) *Trump's Media War.* Palgrave Macmillan, Cham, pp 227–240.

70 Palmer, J. (2019) How does online racism spawn mass shooters? *Foreign Policy*, 4 Aug. https://foreignpolicy.com/2019/08/04/online-racism-4chan-8chan-shootings-elpaso-dayton-texas-ohio/; Glaser, A. (2019) 8chan is a normal part of mass shootings now. *Slate*, 4 Aug. https://slate.com/technology/2019/08/el-paso-8chan-4chan-mass-shootings-manifesto.html.

71 Innes, M. (2001) Control creep. *Sociological Research Online* 6(3), http://www.socresonline.org.uk/6/3/innes.html.

72 Meissner, M. and Wubbeke, J. (2016) IT-backed authoritarianism: Information technology enhances central authority and control capacity under Xi Jinping. *MERICS China Monitor*. https://www.merics.org/sites/default/files/2017-09/MPOC_ChinasCoreExecutive.pdf.

73 Liang, F., Das, V., Kostyuk, N. and Hussain, M. M. (2018) Constructing a data-driven society: China's Social Credit System as a state surveillance infrastructure. *Policy and Internet* 10(4): 415–53.

74 Liang et al (2018).

75 Hoffman, S. (2017) Managing the state: Social credit, surveillance and the CCP's plan for China. *China Brief* 17(11), 17 Aug, https://jamestown.org/program/managing-the-state-social-credit-surveillance-and-the-ccps-plan-for-china/; Liang et al (2018).

76 Meissner, M. and Wubbeke, J. (2016).

77 Chan, K. (2019) Police, protesters clash in Hong Kong as demonstrators cut down smart lamppost. *Global News*, 24 Aug. https://globalnews.ca/news/5808861/hong-kong-protests-smart-lamppost/.

78 In effect, this 'money' is value, 'surplus value' in Marx's terms.

Chapter 4

1 Cheney-Lippold, J. (2017) p 194.

2 Carlo Petrini, founder of the Slow Food movement, cited in Honoré, C. (2005) p 16.

3 Rosa, H. (2015) p 321.

4 Pang, A. S-K. (2016).

5 Honoré, C. (2005).

6 Pang, A. S-K. (2016).

7 Honoré, C. (2005).

8 Honoré, C. (2005).

9 Carter, B., Rees, P. and Hale, L. (2016) Association between portable screen-based media device access or use and sleep outcomes. *JAMA Pediatrics* 170(12): 1202–08. https://jamanetwork.com/journals/jamapediatrics/fullarticle/2571467; Loria, K. and Gould, S. (2017) How smartphone light affects your brain and body. *The Business Insider*, 11 July. https://www.businessinsider.com/how-smartphone-light-affects-your-brain-and-body-2017-7.

10 Pang, A. S-K. (2016).

11 Chen, B. X. (2018).

12 Hayles, K. N. (2017) *Unthought: The power of the cognitive unconscious.* University of Chicago Press, Chicago, p 203.

13 Cheney-Lippold, J. (2016) p 198.

14 Fraser, A. (2019) Land grab/data grab: Precision agriculture and its new horizons, *The Journal of Peasant Studies* 46(5): 893–912.

15 Kukutai, T. and Taylor, J. (eds) (2016) *Indigenous Data Sovereignty: Towards An Agenda.* Australian National University Press, Acton; Mann, M. and Daly, A. (2019) (Big) Data and the North-in-South: Australia's Informational Imperialism and Digital Colonialism. *Television & New Media* 20(4): 379–95; Global Indigenous Data Alliance (GIDA) (2019) *CARE Principles for Indigenous Data Governance.* https://www.gida-global.org/care

16 Cardullo, P. and Kitchin, R. (2019) Being a 'citizen' in the smart city: Up and down the scaffold of smart citizen participation in Dublin, Ireland. *GeoJournal* 84(1): 1–13.

17 This is an assigned ID to enable data marketers to track your use of different apps and push advertising to you. To disable the ad ID: on Android navigate to Settings > Privacy > Advanced > Ads and toggle 'Reset advertising ID' or 'Opt out of ads personalization'; on iOS, go to Settings > Privacy > Advertising and toggle on 'Limit ad tracking' or 'Reset advertising identifier'. Newman, L. H. (2019) A simple way to make it harder for mobile ads to track you. *Wired*, 21 Sept. https://www.wired.com/story/ad-id-ios-android-tracking/.

18 For example, changing settings to turn off the sharing of location, access to contacts or photos, push notifications, or to make sure that features such as the camera and microphone cannot be accessed without specific permission.

19 https://www.mywot.com/; https://safeweb.norton.com; https://global.sitesafety.trendmicro.com/ (it should be noted

that these are also monitoring what websites you visit or are interested in).

20 Such as Better Business Bureau (https://www.bbb.org/); Online Trust Alliance (https://otalliance.org/); and TrustArc (https://www.trustarc.com/).

21 https://www.facebook.com/help/1701730696756992.

22 See https://aboutthedata.com/.

23 Article 17, GDPR, Right to erasure ('right to be forgotten'). https://gdpr-info.eu/art-17-gdpr/.

24 Khandelwal, S. (2017) Google collects Android location data even when location service is disabled. *The Hacker News*, 21 Nov. https://thehackernews.com/2017/11/android-location-tracking.html.

25 What is open source? https://opensource.com/resources/what-open-source.

26 If you are using a corporate computer or smartphone you should check with your IT department about installing open source operating systems, and other open source programs, before doing so.

27 For an extensive list, including sound and writing packages, see Cullen Vance's post: https://www.facebook.com/cullenvance/posts/2111143198907530.

28 Brunton, F. and Nissenbaum, H. (2016) *Obfuscation: A User's Guide for Privacy and Protest*. MIT Press, Cambridge, MA.

29 Cheney-Lippold, J. (2017) p 231.

30 https://cs.nyu.edu/trackmenot/.

31 Cheney-Lippold, J. (2017) p 232.

32 https://adnauseam.io/.

33 Virtual Private Network, Wikipedia. https://en.wikipedia.org/wiki/Virtual_private_network.

34 Goodman, M. (2015) *Future Crimes: A Journey to the Dark Side of Technology – and How to Survive It*. Bantam Press, New York.

35 Cheney-Lippold, J. (2017) p 239.

36 Cheney-Lippold, J. (2017) p 240.

37 Cheney-Lippold, J. (2017) p 241.

38 Honoré, C. (2005).

39 Honoré, C. (2005) p 34.

40 From the book *Postcards from the Edge* by Carrie Fisher (1987), cited in Honoré, C. (2005) p 12.

41 Lustig, R. (2017).

42 Eyal, N. (2014).

43 Lustig, R. (2017).

44 Eyal, N. (2014).
45 Oberhaus, D. (2018) How I quit Apple, Microsoft, Google, Facebook, and Amazon. *Motherboard*, 13 Dec. https://motherboard.vice.com/en_us/article/ev3qw7/how-to-quit-apple-microsoft-google-facebook-amazon.
46 See Desjardins, J. (2018) These are the world's largest tech giants. World Economic Forum. https://www.weforum.org/agenda/2018/07/visualizing-the-world-s-20-largest-tech-giants for the 2018 valuations of these companies.
47 Fisher, M. (2009) *Capitalist Realism: Is There No Alternative?* Zero Books, Ropley, Hants.
48 Dencik, L. (2018) Surveillance realism and the politics of imagination: Is there no alternative? *Krisis: Journal for Contemporary Philosophy*, Issue 1, Data Activism: 36–42.
49 Smith, G. J. D. (2018) Data doxa: The affective consequences of data practices. *Big Data and Society* 5: 1–15.
50 Fisher, M. (2009) p 21.
51 Dodge, M. and Kitchin, R. (2005) Code and the transduction of space. *Annals of the Association of American Geographers* 95(1): 162–180.
52 Dodge, M. and Kitchin, R. (2005) Codes of life: Identification codes and the machine-readable world. *Environment and Planning D: Society and Space* 23(6): 851–81.

Chapter 5

1 Right to Disconnect. Wikipedia. https://en.wikipedia.org/wiki/Right_to_disconnect.
2 Right to Disconnect. Wikipedia. https://en.wikipedia.org/wiki/Right_to_disconnect.
3 Right to Disconnect. Wikipedia. https://en.wikipedia.org/wiki/Right_to_disconnect.
4 Ortega, F. (2009). The cerebral subject and the challenge of neurodiversity. *BioSocieties* 4(4): 425–45.
5 Eurofound (2016).
6 Farnsworth, C. B. (2015) Dutch architect argues for Faraday-like safe rooms that IOT can't penetrate. *Green Builder*, 19 May. https://www.greenbuildermedia.com/Internet-of-things/safe-room-haven-from-iot.
7 Ram House (2015). Space Caviar. http://www.spacecaviar.net/ram-house/.

8   Stinson, L. (2013) This signal-blocking Faraday cage might drive you crazy. *Wired*, 17 Dec. https://www.wired.co.uk/article/life-size-faraday-cage.

9   Manaugh, G. (2015) New urbanist: Giving physical shape to invisible signals. *New Scientist*, 9 June. https://www.newscientist.com/article/dn27685-new-urbanist-giving-physical-shape-to-invisible-signals/.

10 Angwin, J. (2014), p 223.

11 Minelli et al (2013); Mayer-Schonberger and Cukier (2013).

12 Article 29 Data Protection Working (2014); Fuster, G. G. and Scherrer, A. (2015) *Big Data and Smart Devices and Their Impact on Privacy*. Committee on Civil Liberties, Justice and Home Affairs (LIBE), Directorate-General for Internal Policies, European Parliament. http://www.europarl.europa.eu/RegData/etudes/STUD/2015/536455/IPOL_STU(2015)536455_EN.pdf.

13 Parakilas, S. (2017) We can't trust Facebook to regulate itself. *New York Times*, 19 Nov. https://www.nytimes.com/2017/11/19/opinion/facebook-regulation-incentive.html.

14 Dance, G. J. X., Confessore, N. and LaForgia, M. (2018) Facebook gave device makers deep access to data on users and friends. *New York Times,* 3 June. https://www.nytimes.com/interactive/2018/06/03/technology/facebook-device-partners-users-friends-data.html.

15 Newton, C. (2018) Google's new focus on well-being started five years ago with this presentation. *The Verge*, 10 May. https://www.theverge.com/platform/amp/2018/5/10/17333574/google-android-p-update-tristan-harris-design-ethics.

16 Center for Humane Technology. http://humanetech.com/.

17 Google Digital Wellbeing. https://wellbeing.google/.

18 Angwin, J. (2014).

19 Hern, A. (2018) No tracking, no revenue: Apple's privacy feature costs ad companies millions. *The Guardian*, 9 Jan. https://www.theguardian.com/technology/2018/jan/09/apple-tracking-block-costs-advertising-companies-millions-dollars-criteo-web-browser-safari.

20 Corbyn, Z. (2018) Decentralisation: The next big step for the world wide web. *The Guardian*, 8 Sept. https://www.theguardian.com/technology/2018/sep/08/decentralisation-next-big-step-for-the-world-wide-web-dweb-data-internet-censorship-brewster-kahle.

21 Corbyn, Z. (2018).
22 Harbinja, E. and Karagiannopoulos, V. (2019) Web 3.0: The decentralised web promises to make the internet free again. *The Conversation*, 11 Mar. https://theconversation.com/web-3-0-the-decentralised-web-promises-to-make-the-internet-free-again-113139.
23 As Edina Harbinja and Vasileios Karagiannopoulos (2019) note: 'Some of the technologies that could make the DWeb possible are already being developed. For example, the Databox Project (https://www.databoxproject.uk/about/) aims to create an open source device that stores and controls a user's personal data locally instead of letting tech companies gather and do whatever they like with it. ZeroNet (https://zeronet.io/) is an alternative to the existing web where websites are hosted by a network of participating computers instead of a centralised server, protected by the same cryptography that's used for bitcoin. There's even a DWeb version of YouTube called DTube (https://about.d.tube/), which hosts videos across a decentralised network of computers using a blockchain-based public ledger as its database and payment system.'
24 See Barabas, C. et al (2017) *Defending Internet Freedom through Decentralization: Back to the Future?* http://dci.mit.edu/assets/papers/decentralized_web.pdf.
25 Meyer, D. (2018) Telegram starts to play nice with security agencies over user data, but not in Russia. ZDNet, 29 Aug. https://www.zdnet.com/article/telegram-starts-to-play-nice-with-security-agencies-over-user-data-but-not-in-russia/.
26 See Minelli et al (2013) p 156; OECD (1980) *OECD Guidelines on the Protection of Privacy and Transborder Flows of Personal Data.* http://www.oecd.org/sti/ieconomy/oecdguidelinesontheprotectionofprivacyandtransborderflowsofpersonaldata.htm.
27 The White House (2012) Consumer data privacy in a networked world: A framework of protecting privacy and promoting innovation in the global digital economy. http://www.whitehouse.gov/sites/default/files/privacy-final.pdf; Article 29 Data Protection Working (2014); FTC (2000) *Privacy Online: Fair Information Practice Principles in the Electronic Marketplace.* Federal Trade Commission, Washington DC. http://www.ftc.gov/sites/default/files/documents/reports/privacy-online-fair-information-practices-electronic-marketplace-federal-trade-commission-report/privacy2000text.pdf; Ramirez (2013); Lomas, N. (2015) The FTC warns Internet of Things

businesses to bake in privacy and security. *TechCrunch*, 8 Jan. http://techcrunch.com/2015/01/08/ftc-iot-privacy-warning; New Zealand Data Futures Forum (2014) *Harnessing the Social and Economic Power of Data*. http://www.nzdatafutures.org.nz/sites/default/files/NZDFF_harness-the-power.pdf; CIPPIC (2006).

28 S. 1158 (114th): Consumer Privacy Protection Act of 2015. GovTrack. https://www.govtrack.us/congress/bills/114/s1158.

29 Cavoukian, A. (2009).

30 Cavoukian, A. and Castro, D. (2014) *Big Data and Innovation, Setting the Record Straight: De-identification Does Work*. Information and Privacy Commissioner Ontario, Canada. http://www2.itif.org/2014-big-data-deidentification.pdf.

31 Davis, B. (2017) GDPR requires privacy by design, but what is it and how can marketers comply? *Econsultancy*, 25 Aug. https://econsultancy.com/gdpr-requires-privacy-by-design-but-what-is-it-and-how-can-marketers-comply/.

32 Carson, A. (2014) Seattle launches sweeping, ethics-based privacy overhaul. *The Privacy Advisor*, 7 Nov; Goldsmith, S. (2015) Protecting big data: Seattle's digital privacy initiative aims to keep innovation on track with new data safeguards. *Data-Smart City Solutions*, 29 Sept. https:// datasmart.ash. harvard.edu/news/article/protecting-big-data-742.

33 http://www.seattle.gov/information-technology/privacy-program.

34 Ettlinger, N. (2018) Algorithmic affordances for productive resistance. *Big Data and Society* 5: 1–13.

35 Milan, S. and van der Velden, L. (2016) The alternative epistemologies of data activism. *Digital Culture & Society* 2(2): 57–74.

36 Milan, S. and van der Velden, L. (2016).

37 Schrock, A. (2018) *Civic Tech*. Rogue Academic Press, Los Angeles.

38 Powell, A. (2008) Wifi publics: Producing community and technology. *Information, Communication & Society* 11(8): 1068–88; Cardullo, P. (2017) Gentrification in the mesh? An ethnography of Open Wireless Network (OWN) in Deptford. *City* 21(3-4): 405–19.

39 Tucker, B. (2017) How a remote South African rural community, with barely any electricity, built its own ISP. *Quartz*

Africa, 9 Dec. https://qz.com/africa/1152288/an-eastern-cape-rural-community-in-south-africa-have-their-own-isp/.

40 King, J. (2012) A tech innovation in Detroit: Connect people, not computers. *ColorLines*, 3 Oct. https://www.colorlines.com/articles/tech-innovation-detroit-connect-people-not-computers.

41 Lepp, H. (2015) An investigation of decentralized networks based upon wireless mobile technologies. *Intersect* 8(2): 1–16.

42 Meyer, R. (2014) What Firechat's success in Hong Kong means for a global internet. *The Atlantic*, 6 Oct. https://www.theatlantic.com/technology/archive/2014/10/firechat-the-hong-kong-protest-tool-aims-to-connect-the-next-billion/381113/.

43 Lepp, H. (2015).

44 Platform Cooperativism Consortium. https://platform.coop/.

45 Fraser, A. (2019).

46 Grower Data Coop. https://www.gisc.coop/.

47 AgXchange. https://agproexchange.com/about/.

48 Dodge, M. and Kitchin, R. (2013) Crowdsourced cartography: Mapping experience and knowledge. *Environment and Planning A* 45(1): 19–36.

49 Perng, S-Y., Kitchin, R. and MacDonncha, D. (2018) Hackathons, entrepreneurial life and the making of smart cities. *Geoforum* 97: 189–197.

50 #Hack4Good. https://www.hack4good.io/.

51 Data for Black Lives. http://d4bl.org/.

52 Detroit Digital Justice Coalition. http://detroitdjc.org/data-justice/.

53 'Promoting Data Protection by Privacy Enhancing Technologies (PETs)', COM(2007) 228 final, cited in European Data Protection Supervisor (2014).

54 Privacy Enhancing Technologies. Wikipedia. https://en.wikipedia.org/wiki/Privacy-enhancing_technologies.

55 See The Center for Internet and Society. PET. https://cyberlaw.stanford.edu/wiki/index.php/PET for a list of PETs.

56 A Global Standard for Data Protection Law. Privacy International. https://privacyinternational.org/impact/global-standard-data-protection-law.

57 EDRi (2018) GDPR: A new philosophy of respect. EDRi, 24 May.https://edri.org/press-release-gdpr-philosophy-respect/.

58 Jarvinen, H. (2015) EU Data Protection Package – Lacking ambition but saving the basics. EDRi, 17 Dec. https://edri.

org/eu-data-protection-package-lacking-ambition-but-saving-
the-basics/.

59   Liberty (2018) Liberty wins first battle in landmark challenge
to mass surveillance powers in the Investigatory Powers Act.
Liberty, 27 Apr. https://www.libertyhumanrights.org.uk/
news/press-releases-and-statements/liberty-wins-first-battle-
landmark-challenge-mass-surveillance.

60   McGuire, D. (2018) Connecticut's plan to install electronic
tolling could be a privacy nightmare. ACLU, 25 July. https://
www.aclu.org/blog/privacy-technology/location-tracking/
connecticuts-plan-install-electronic-tolling-could-be.

61   Bitar, J. and Stanley, J. (2018) Are stores you shop at secretly
using face recognition on you? ACLU, 26 Mar. https://www.
aclu.org/blog/privacy-technology/surveillance-technologies/
are-stores-you-shop-secretly-using-face.

62   Handeyside, H. (2018) We're demanding the government come
clean on surveillance of social media. ACLU, 24 May. https://
www.aclu.org/blog/privacy-technology/Internet-privacy/
were-demanding-government-come-clean-surveillance-social.

Chapter 6

1   Held, V. (2005) *The Ethics of Care*. Oxford University Press,
Oxford; Tronto, J. C. (1993) *Moral Boundaries: A Political
Argument for an Ethic of Care*. Routledge, New York.

2   Tronto, J. C. (2005) An ethic of care. In Cudd, A. E. and
Andreasen, R. O. (eds) *Feminist Theory: A Philosophical
Anthology*. Blackwell, Oxford, pp 251–63.

3   For explanations of different theories of social justice, see:
Sabbagh, C. and Schmitt, M. (eds) (2016) *Handbook of Social
Justice Theory and Research*. Springer, New York; Smith, D. M.
(1994) *Geography and Social Justice*. Blackwell, Oxford.

4   Gilligan, C. (1987) Moral orientation and moral development.
In Kittay, E. and Meyers, D. (eds) *Women and Moral Theory*.
Rowman & Littlefield Publishers, Totowa, NJ, pp 19–33.

5   Tronto, J. C. (2005).

6   Eichstaedt, P. (2011) *Consuming the Congo: War and Conflict
Minerals in the World's Deadliest Place*. Lawrence Hill Books,
Chicago; Fowler, B. A. (2018) *Electronic Waste: Toxicology and
Public Health Issues*. Academic Press, London; Lepawsky, J.
(2018) *Reassembling Rubbish: Worlding Electronic Waste*. MIT
Press, Cambridge, MA.

7   Tarnoff, B. (2019) To decarbonize we must decomputerize:
      Why we need a Luddite revolution. *The Guardian*, 18 Sept.
      https://www.theguardian.com/technology/2019/sep/17/
      tech-climate-change-luddites-data.
8   Right to Disconnect. Wikipedia. https://en.wikipedia.org/
      wiki/Right_to_disconnect.
9   Delmas, M.A. and Burbano, V. C. (2011) The drivers of
      greenwashing. *California Management Review* 54(1): 64–87.
10   Wagner, B. (2018) Ethics as an escape from regulation: From
      ethics-washing to ethics-shopping? In Hildebrandt, M. et al
      (eds) *Being Profiling: Cogitas Ergo Sum*. Amsterdam University
      Press, Amsterdam, pp 84–89; O'Keefe, K. and O'Brien, D.
      (2018) *New Paradigm or Ethics Washing? An Analysis of Facebook's
      Ethics Report*. Castlebridge, Dublin. https://www.castlebridge.
      ie/new-paradigm-or-ethics-washing-an-analysis-of-facebooks-
      ethics-report/.
11   Parsons, W. (2004) Not just steering but weaving: Relevant
      knowledge and the craft of building policy capacity and
      coherence. *Australian Journal of Public Administration* 63(1):
      43–57; Kitchin, R., Lauriault, T. and McArdle, G. (2015)
      Knowing and governing cities through urban indicators, city
      benchmarking and real-time dashboards. *Regional Studies,
      Regional Science* 2: 1–28.
12   Leccardi, C. (2007); Hassan, R. (2007); Bleecker, J. and Nova,
      N. (2009); de Lange, M. (2018).
13   Bleecker, J. and Nova, N. (2009).
14   Bleecker, J. and Nova, N. (2009).
15   Adams, B. and Groves, C. (2007) *Future Matters: Action,
      Knowledge, Ethics*. Brill, Leiden.
16   For example, we can forecast tides and lunar cycles quite
      accurately, and the weather quite well.
17   Seligman et al (2016).
18   Adams, B. and Grove, C. (2007).
19   Söderström, O., Paasche, T., and Klauser, F. (2014) Smart cities
      as corporate storytelling. *City*, 18(3): 307–20; Sadowski, J. and
      Bendor, R. (2019) Selling smartness: Corporate narratives and
      the smart city as a sociotechnical imaginary. *Science, Technology,
      & Human Values*, 44(3): 540–563.
20   Datta, A. (2016) Introduction: Fast cities in an urban age. In
      Datta, A. and Shaban, A. (eds) *Mega-Urbanization in the Global
      South: Fast Cities and New Urban Utopias of the Postcolonial State*.
      Routledge, London, pp 1–27.

21 Datta, A. and Shaban, A. (2016) Slow: Towards a decelerated urbanism. In Datta, A. and Shaban, A. (eds) *Mega-Urbanization in the Global South: Fast Cities and New Urban Utopias of the Postcolonial State*. Routledge, London, pp 205–20; Kitchin, R. (2019) Towards a genuinely humanizing smart urbanism. In Cardullo, P., di Feliciantonio, C. and Kitchin, R. (eds) *The Right to the Smart City*. Emerald, Bingley, pp 193–204.

22 Kitchin, R., Cardullo, P. and di Feliciantonio, C. (2019).

23 Adam, B. (2008) Of timespaces, futurescapes and timeprints. Presentation at Lüneburg University. http://www.cardiff.ac.uk/socsi/futures/conf_ba_lueneberg170608.pdf.

24 Anderson, B. (2010) Preemption, precaution, preparedness: Anticipatory action and future geographies. *Progress in Human Geography* 34(6): 777–98.

25 Adams, B. and Grove, C. (2007).

26 Groves, C. (2009) Future ethics: Risk, care and non-reciprocal responsibility. *Journal of Global Ethics* 5(1): 17–31.

27 For introduction to data justice see the websites of four ongoing projects: Data Justice Lab at Cardiff University: https://datajusticelab.org/; Global Data Justice at Tilberg University: https://globaldatajustice.org/; Datactive at Amsterdam University: https://data-activism.net/; and Our Data Bodies in the United States: https://www.odbproject.org/.

28 Dencik, L., Hintz, A. and Cable, J. (2016) Towards data justice? The ambiguity of anti-surveillance resistance in political activism. *Big Data & Society* 3(2): 1–12; Taylor, L. (2017) What is data justice? The case for connecting digital rights and freedoms globally. *Big Data & Society* 4(2), July–Dec: 1–14.

29 Sabbagh, C. and Schmitt, M. (2016).

30 Santucci, G. (2013).

31 Tarin, D. (2015) Privacy and Big Data in smart cities. *AC Actual Smart City*, 28 Jan. http://www.smartcities.com/en/latest3/tech-2/item/503-privacy-and-big-data-in-smart-cities.

32 Schneier, B. (2015) How we sold our souls – and more – to the Internet giants. *The Guardian*, 17 May. www.theguardian.com/technology/2015/may/17/sold-our-souls-and-more-to-Internet-giants-privacy-surveillance-bruce-schneier.

33 Cohen, J. (2012); Cavoukian, A. (2009).

34 Rambam, S. (2008); Rubenking, N. (2013).

35 Cohen, J. (2012); Solove, D. J. (2007).

36 Schneier, B. (2015).

37   As Angwin's (2014) concerted attempts to reclaim her privacy highlight, at present it is very difficult to regain any meaningful level of protection or redress with respect to the mass generation of big data. Despite being technically savvy and having access to leading experts in the field she struggled to find technical solutions that limited the data generated about her.
38   Dodge, M. and Kitchin, R. (2007).
39   Schacter, D. L. (2001) *The Seven Sins of Memory: How the Mind Forgets and Remembers.* Houghton Mifflin, Boston, MA.
40   Kitchin, R. (2014).
41   Bollier, D. (2010) *The Promise and Peril of Big Data.* The Aspen Institute. http://www.aspeninstitute.org/sites/default/files/content/docs/pubs/The_Promise_and_Peril_of_Big_Data.pdf.
42   Kitchin, R. (2017) Thinking critically about and researching algorithms. *Information, Communication and Society* 20(1): 14–29.
43   Kitchin, R. (2017).
44   Seaver, N. (2013) Knowing algorithms, in *Media in Transition 8*, Cambridge, MA. http://nickseaver.net/papers/seaverMiT8.pdf.
45   Montfort, N., Baudoin, P., Bell, J., Bogost, I., Douglass, J., Marino, M. C., Mateas, M., Reas, C., Sample, M. and Vawter, N. (2012) *10 PRINT CHR$ (205.5 + RND (1)); : GOTO 10.* MIT Press, Cambridge, MA.
46   Kitchin, R. and Dodge, M. (2011).
47   Gillespie, T. (2014a). The relevance of algorithms. In Gillespie, T., Boczkowski, P. J. and Foot, K. A. (eds) *Media Technologies: Essays on Communication, Materiality, and Society.* MIT Press, Cambridge, MA. pp 167–93.
48   Taylor, L. (2017).
49   Bostrom, N. and Yudkowsky, E. (2014) The ethics of artificial intelligence. In Frankish, K. and Ramsey, W. (eds) *The Cambridge Handbook of Artificial Intelligence.* Cambridge University Press, Cambridge, pp 316–34.
50   There is an emerging debate within academia and institutional circles concerning AI and ethics. For example, the EU has established a High-Level Expert Group on AI that will lead to a European strategy on Artificial Intelligence to include recommendations on ethical, legal, and societal issues related to AI. See https://ec.europa.eu/digital-single-market/en/high-level-expert-group-artificial-intelligence.
51   Honore, C. (2005) p 49.
52   Wajcman, J. (2015).

53 Woodcock, J. and Graham, M. (2019) *The Gig Economy: A Critical Introduction*. Polity Books, Cambridge.
54 Johnson, J. A. (2013) From open data to information justice. Paper presented at the Annual Conference of the Midwest Political Science Association, 13 April, Chicago, Illinois. http://papers.ssrn.com/abstract=2241092.
55 Eubanks, V. (2018).
56 Benjamin, R. (2019).
57 Stroud, M. (2014); Edwards, E. (2016) Predictive policing software is more accurate at predicting policing than predicting crime. *Huffington Post*, 31 Aug. https://www. huffingtonpost.com/entry/predictive-policing-reform_ us_57c6ffe0e4b0e60d31dc9120; Jefferson, B. J. (2018) Policing, data, and power-geometry: Intersections of crime analytics and race in urban restructuring. *Urban Geography* 39(8): 1247–64.
58 Harcourt, B. (2008); Benjamin, R. (2019).
59 Noble, S. U. (2018).
60 Vincent, J. (2016) Twitter taught Microsoft's AI chatbot to be a racist asshole in less than a day. *The Verge*, 24 Mar. https:// www.theverge.com/2016/3/24/11297050/tay-microsoft-chatbot-racist.
61 Benjamin, R. (2019).
62 Petty, T., Saba, M., Lewis, T., Gangadharan, S. P. and Eubanks, V. (2016) *Our Data Bodies: Reclaiming Our Data*. Our Data Bodies Project. https://www.odbproject.org/wp-content/ uploads/2016/12/ODB.InterimReport.FINAL_.7.16.2018. pdf.
63 Petty, T., et al (2016).

Chapter 7

1 Drake, J. D. (2001) *Downshifting: How to Work Less and Enjoy Life More*. Berrett-Koehler Publishers, Oakland, CA.; Ghazi, P. and Jones, J. (2004) *Downshifting: A Guide to Happier, Simpler Living*. Hodder & Stoughton, London.
2 Jones, J. (2017) George Orwell predicted cameras would watch us in our homes; he never imagined we'd gladly buy and install them ourselves. *Open Culture*, 28 Nov. http://www. openculture.com/2017/11/george-orwell-never-imagined-wed-gladly-buy-and-install-cameras-in-our-homes.html.
3 Keith Lowell Jensen (2013) https://twitter.com/keithlowell/ status/347741181997879297?lang=en.

4   For example, the mandatory replacement of utility meters with smart meters that provide a much more fine-grained picture of the use of electricity, gas, and water in a home.

5   For example, Uber actively lobbies at city and national level around the world for deregulation of the taxi industry to enable it to legitimize its move into new markets.

Coda

1   *The Economist* (2020) Countries are using apps and data networks to keep tabs on the pandemic. *The Economist*, 26 March. https://www.economist.com/briefing/2020/03/26/countries-are-using-apps-and-data-networks-to-keep-tabs-on-the-pandemic

2   Kitchin, R. (2020) Using digital technologies to tackle the spread of the coronavirus: Panacea or folly? *Programmable City Working Paper 44*. See http://progcity.maynoothuniversity.ie/wp-content/uploads/2020/04/Digital-tech-spread-of-coronavirus-Rob-Kitchin-PC-WP44.pdf

3   Goh, B. (2020) China rolls out fresh data collection campaign to combat coronavirus. *Reuters*, 26 February. See https://www.reuters.com/article/us-china-health-data-collection/china-rolls-out-fresh-data-collection-campaign-to-combat-coronavirus-idUSKCN20K0LW

4   Nielsen, M. (2020) Privacy issues arise as governments track virus. *EU Observer*, 23 March. See https://euobserver.com/coronavirus/147828

5   Cahane, A. (2020) The Israeli emergency regulations for location tracking of coronavirus carriers. *Lawfare*, 21 March. See https://www.lawfareblog.com/israeli-emergency-regulations-location-tracking-coronavirus-carriers

6   Linklaters (2020) 28 countries race to launch official Covid-19 tracking apps to reduce the spread of the virus. 16 April. See https://www.linklaters.com/en/about-us/news-and-deals/deals/2020/april/28-countries-race-to-launch-official-covid-19-tracking-apps-to-reduce-the-spread-of-the-virus

7   Nellis, S. (2020) As fever checks become the norm in coronavirus era, demand for thermal cameras soars. *Reuters*, 9 April. See https://www.reuters.com/article/us-health-coronavirus-thermal-cameras-fo/as-fever-checks-become-the-norm-in-coronavirus-era-demand-for-thermal-cameras-soars-idUSKCN21R2SF; Url, S. (2020) Drones take Italians'

temperature and issue fines. *Arab News*, 10 April. See https://www.arabnews.com/node/1656576/world

8   Brandom, R. and Robertson, A. (2020) Apple and Google are building a coronavirus tracking system into iOS and Android. *The Verge*, 10 April. See https://www.theverge.com/2020/4/10/21216484/google-apple-coronavirus-contract-tracing-bluetooth-location-tracking-data-app

9   Pollina, E. and Busvine, D. (2020) European mobile operators share data for coronavirus fight. *Reuters*, 18 March. See https://www.reuters.com/article/us-health-coronavirus-europe-telecoms/european-mobile-operators-share-data-for-coronavirus-fight-idUSKBN2152C2

10 Fowler, G.A. (2020) Smartphone data reveal which Americans are social distancing (and not). *Washington Post*, 24 March. See https://www.washingtonpost.com/technology/2020/03/24/social-distancing-maps-cellphone-location/

11 Wodinsky, S. (2020) Experian is tracking the people most likely to get screwed over by coronavirus. *Gizmodo*, 15 April. See https://gizmodo.com/experian-is-tracking-the-people-most-likely-to-get-scre-1842843363

12 Mosendz, P. and Melin, A. (2020) Bosses are panic-buying spy software to keep tabs on remote workers. *Los Angeles Times*, 27 March. See https://www.latimes.com/business/technology/story/2020-03-27/coronavirus-work-from-home-privacy

13 Singer, N. and Sang-Hun, C. (2020) As coronavirus surveillance escalates, personal privacy plummets. *New York Times*, 23 March. See https://www.nytimes.com/2020/03/23/technology/coronavirus-surveillance-tracking-privacy.html

14 Kitchin, R. (2020); French, M. and Monahan, T. (2020) Dis-ease surveillance: how might surveillance studies address COVID-19? *Surveillance Studies* 18(1). See https://ojs.library.queensu.ca/index.php/surveillance-and-society/article/view/13985

15 Guarglia, M. and Schwartz, A. (2020) *Protecting civil liberties during a public health crisis*. Electronic Frontier Foundation, 10 March. See https://www.eff.org/deeplinks/2020/03/protecting-civil-liberties-during-public-health-crisis

16 Stanley, J. and Granick, J.S. (2020) *The limits of location tracking in an epidemic*. ACLU, 8 April. See https://www.aclu.org/sites/default/files/field_document/limits_of_location_tracking_in_an_epidemic.pdf

17 Ada Lovelace Institute (2020) *Exit Through The App Store.* 20
    April. See https://www.adalovelaceinstitute.org/wp-content/
    uploads/2020/04/Ada-Lovelace-Institute-Rapid-Evidence-
    Review-Exit-through-the-App-Store-April-2020-1.pdf
18 European Data Protection Board (2020) *Guidelines 04/2020
    on the use of location data and contact tracing tools in the context of
    the COVID-19 outbreak.* 21 April. See https://edpb.europa.
    eu/sites/edpb/files/files/file1/edpb_guidelines_20200420_
    contact_tracing_covid_with_annex_en.pdf

Index

time
  clock 31–2, 33, 35, 79, 89, 138,
    152, 161
  dead 22, 138, 153, 166
  just-in- 32, 40
  network 32, 35, 40, 79, 138, 140,
    152, 161, 167
  pressure 2, 3, 10, 11, 12, 13, 14,
    37, 38, 39, 41, 43, 45, 77, 78, 81,
    98, 99, 158, 159, 165, 169, 175
  scarcity 17, 39, 42, 45
  shifting 22, 32, 138, 152, 159
  sovereignty 77, 78–84, 137–44,
    159, 161, 167
  stress 2, 6, 10, 12, 13, 17, 21, 23,
    39, 47, 79, 107, 110, 137, 138,
    139, 157, 158, 159, 162, 170, 171
time-space compression 23, 25–32,
  35, 36, 37
Tinder 65, 94
Tor 97–8, 109, 127, 163
transparency 11, 60, 91, 92, 150,
  174
transponder 31, 56
travel
  air 6, 27–8, 33, 59, 148
  car 5, 23, 25, 26, 27, 33, 46, 53, 56,
    106, 129
  rail 23, 26, 27, 33
  ship 26–7
Twitter 1, 20, 30, 34, 58, 61, 81, 83,
  94, 102, 132, 149, 154

U
Uber 8, 45, 94, 152
unions 22, 109, 112, 113, 128, 140,
  153, 165
United Kingdom 1, 9, 21, 30
United States 1, 6, 22, 27, 28, 29,
  39, 51, 65, 67, 71–2, 123, 131,
  154, 155

V
video 3, 29, 30, 33, 34, 35, 61, 88,
  95, 141, 162, 170, 173
virtual private network 97, 163
Volkswagen 110, 112

W
Web 2.0 30
well-being 12, 17, 41, 60, 61,
  79–80, 105, 119, 136, 138, 139,
  140 145, 159, 174
WhatsApp 1, 10, 38, 44, 45, 56, 76,
  81, 94, 170
WiFi 2, 20, 30, 33, 50, 55, 63, 80,
  82, 88, 97, 113, 114, 126, 239,
  161, 163
Wikileaks 9
Wikipedia 30, 128
World Wide Web 30, 119
work 5, 6, 10, 12, 15, 16, 17, 20,
  21, 22, 24, 26, 33, 35, 38, 39, 40,
  42, 50, 57, 68, 69, 72–4, 76, 80,
  81, 82, 89, 99, 109–10, 118, 136,
  140, 151, 152, 159, 160, 166, 171,
  173, 175
  culture 22, 42, 99
  flow 24, 39, 81, 111
  gig 153
  hours 15, 21, 32, 41, 80, 82, 111,
    113, 137, 138, 153, 175
  load 24
  stressed 13
work-life balance 21, 109, 138, 170
workaholic 21, 99, 138
working time drift 21, 46, 79, 109,
  111, 138
workplace 24, 33, 41, 68, 69, 81,
  109, 110, 111, 112, 113, 114,
  170, 173

Y
YouTube 2, 30, 34, 54, 61

Z
Zenzeleni project 126